PACKAGING DESIGN 4

PDC
GOLD AWARDS
Ten Years Of Excellence In Packaging

PDC
PACKAGE
DESIGN
COUNCIL
INTERNATIONAL

By Charles Biondo

PBC INTERNATIONAL ▪ New York

DISTRIBUTOR TO THE BOOK TRADE IN THE UNITED STATES AND CANADA:
Rizzoli International Publications, Inc.
597 Fifth Avenue
New York, NY 10017

DISTRIBUTOR TO THE ART TRADE IN THE UNITED STATES:
Letraset USA
40 Eisenhower Drive
Paramus, NJ 07653

DISTRIBUTOR TO THE ART TRADE IN CANADA:
Letraset Canada Limited
555 Alden Road
Markham, Ontario L3R 3L5, Canada

DISTRIBUTED THROUGHOUT THE REST OF THE WORLD BY:
Hearst Books International
105 Madison Avenue
New York, NY 10016

Copyright © 1988 by PBC INTERNATIONAL, INC.
All rights reserved. No part of this book may be
reproduced in any form whatsoever without written
permission of the copyright owner.

PBC INTERNATIONAL, INC.
One School Street
Glen Cove, NY 11542.

Library of Congress Cataloging-in-Publication Data

Biondo, Charles
    Packaging design 4.

    Includes Index.
    1. Packaging—design.   I. Packaging Design
Council (New York, N.Y. : 1983)   II. Title.
III. Title: Packaging design four.
TS195.2.B56   1988   688.8   88-18030
ISBN 0-86636-065-4

Color separation, printing and binding by
Toppan Printing Co. (H.K.) Ltd. Hong Kong
Typography by **RMP Publication Services**

PRINTED IN HONG KONG
10 9 8 7 6 5 4 3 2 1

CREDITS

Chairman **Charles Biondo**, FPDC
Charles Biondo, Design Associates, NY

Committee Members

John S. Blyth, FPDC/Peterson Blyth Assoc., NY
Ronald Peterson, PDC/Peterson Blyth Assoc., NY
Herbert Meyers, PDC/Gerstman+Meyers, NY
Richard Gerstman, PDC/Gerstman+Meyers, NY

Executive Director of the Competition

Lea Liu, Charles Biondo Design Assoc., NY

STAFF

Managing Director:	PENNY SIBAL-SAMONTE
Financial Director:	PAMELA McCORMICK
Creative Director:	RICHARD LIU
Associate Art Director:	DANIEL KOUW
Editorial Manager:	KEVIN CLARK
Artists:	KIM McCORMICK
	WILLIAM MACK

CONTENTS

Foreword 8

Introduction 10

CHAPTER **1**

Food 12

CHAPTER **2**

Beverages 86

CHAPTER **3**

Health and Beauty Aids 122

CHAPTER **4**

Housewares 154

CHAPTER **5**

Sports and Leisure 166

CHAPTER **6**

Hardware 180

CHAPTER **9**

Corporate Identity 228

CHAPTER **7**

Retail and Soft Goods/Department Store Promotions 204

Index 252

CHAPTER **8**

Office Supplies 216

FOREWORD

There is an unspoken language we all understand.

It is a silent tongue that speaks eloquently of life; a mute voice we hear not merely with our ears but also and primarily through our senses of touch, smell, taste and sight.

This language uses a vocabulary of paper, glass, metal and plastic materials and a rich syntax of textures, colors, shapes and sizes to identify, protect, dispense and market every product produced by man and a host of those bestowed to us by nature.

It is an evocative tongue that can appeal to our emotions without ever uttering a sound. It is a language that sets moods, triggers impulses, satisfies hungers and cares for our health. It is at once prosaic and arcane. It can make us laugh. It gives us life's most precious commodity—time.

The history of this language predates any alphabet. Its contemporary lexicon is a testament to its power, expanding daily to meet our ever-changing needs. Its future—far into the millennia ahead—is more certain than the survival of our species itself.

What is this language, this tongue that speaks to us all in native idioms, as though the Tower of Babbel had never been conceived nor punished?

It is the language of packaging.

And nowhere is the international and universal language of packaging more eloquently articulated than in the Package Design Council International's annual competition. Here we see gathered from the United States, Japan and Europe containers whose elocution and purpose is well understood and appreciated even when the copy carried is in some cryptic dialect we are too linguistically feeble to comprehend.

Pause then, as you page through this book, to appreciate the skills of the talented men and women who so ingeniously crafted these containers, and consider the wonder that is packaging, for at its best, it is a language that makes us all at least bilingual.

— Ben Miyares
Executive Editor
FOOD & DRUG PACKAGING
August 1, 1988

INTRODUCTION

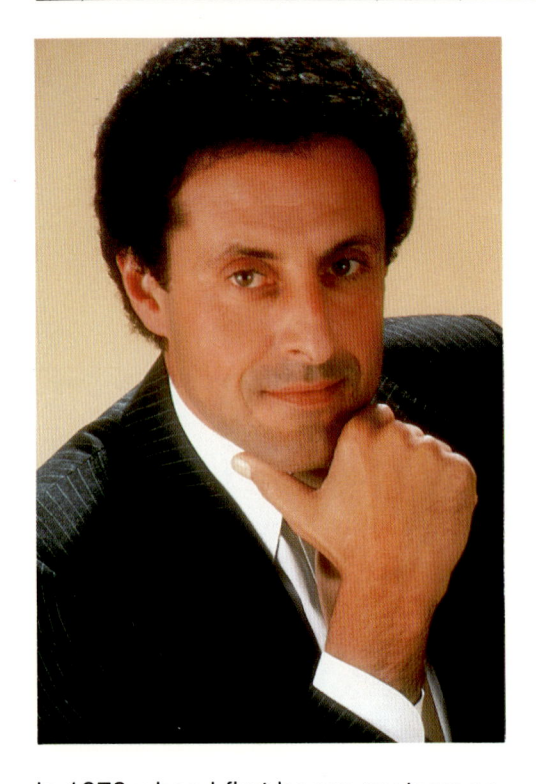

In 1978 when I first began my term as President of PDC, I heard from my colleagues that there was a lack of good junior designers. It occurred to me that at least one good reason for this, might be that some of the best talents were not attending school because of a lack of funds. The prospects for the future development of American package design appeared rather bleak, for all the technical glitter that was then beginning to come on the scene. Clearly, the future of the profession is not an engineering issue.

Perversely, in a society such as ours, the future of the creative mind and the graphic ability which gives expression to it is not limited by restrictions of artistic freedom, but rather, by the inequities which characterize our economic system. I remember saying to our Board of Directors, "We ought to put our money where our mouth is," in a reference to the plight of talented but penniless young designers. This was the simple genesis of the PDC Gold Awards.

It seemed then, and it still does today, that the most appropriate form to raise funds for scholarships, is a competition of professional package design: not only are the proceeds beneficial to young talent yet to be trained, a competition also provides the opportunity for "real life" packages to be measured against each other and the highest standards of the profession.

The panel of judges is carefully selected by the PDC Board of Directors and represents the spectrum of expertise required of packaging: marketing executives, corporate and independent designers, as well as editorial and marketing research professionals. The judges invited by the PDC Board of Directors are recognized to be among the best in their respective fields, and neither substitutions nor offers to judge are accepted.

It is my great pleasure to present to you not only the best of past competitions, but also a glimpse of the discriminating creative minds which have been responsible for the selection of the PDC Gold Awards. In this rare collection you will find the writings of many of the jurors on the outstanding features of various Gold Awards, and among the finalists you will find some of the most eloquent designs of the past decade.

Obviously, these Gold Awards represent only the best of what has been submitted; there are innumerable great packages that have never been entered. Therefore, I hope that through the publication of this book those who have not previously participated will be encouraged to do so, and those who have, will continue to do so. After all, the glory of the PDC Gold Award is transient. Whereas, the endowment of the PDC Scholarship Fund through participation, makes no differentiation between the spirit and intention of the competition: everyone wins. And, that is the point of it all.

I would like to thank all the participants of these PDC Gold Award Competitions for their generous support of the PDC Scholarship Fund, and in particular, to the judges for their dedication in the giving of whole days to the careful evaluation of each entry. I especially would like to thank the members of the Competition Committee, my staff, and the Executive Director of the PDC Competitions, Lea Liu, who has run this competition since its inception. Lastly, I extend my thanks to our publishers for their kind assistance and solidarity with the profession in undertaking this work.

Cordially,

Charles Biondo, FPDC
Chairman
PDC Competition Committee

RONZONI
MARGHERITE
Pasta Primavera
CUT MACARONI•5
PASTA
SMALL SHELLS•23
PASTA
LINGUINE•17
PASTA

Weight Watchers
DELUXE COMBINATION PIZZA
SOUTHERN FRIED CHICKEN PATTY
BEEF ORIENTAL
FILLET OF FISH AU GRATIN
with broccoli

THE STEAMER
The Shower Massage

VISIONS
TOP OF THE RANGE COOKWARE BY CORNING

8"/20 cm SQUARE CAKE DISH
IN RATTAN BASKET
NEW
IDEAL FOR MICROWAVE
pyrex DESIGNS
OVENWARE WITH FLAIR BY CORNING

SHOW IT ALL
BY CORNING

Riché
FRENCH STYLE LOWFAT YOGURT

Impromptu
Spaghetti and Meatballs with Tomato Sauce

CHAPTER 1

Food

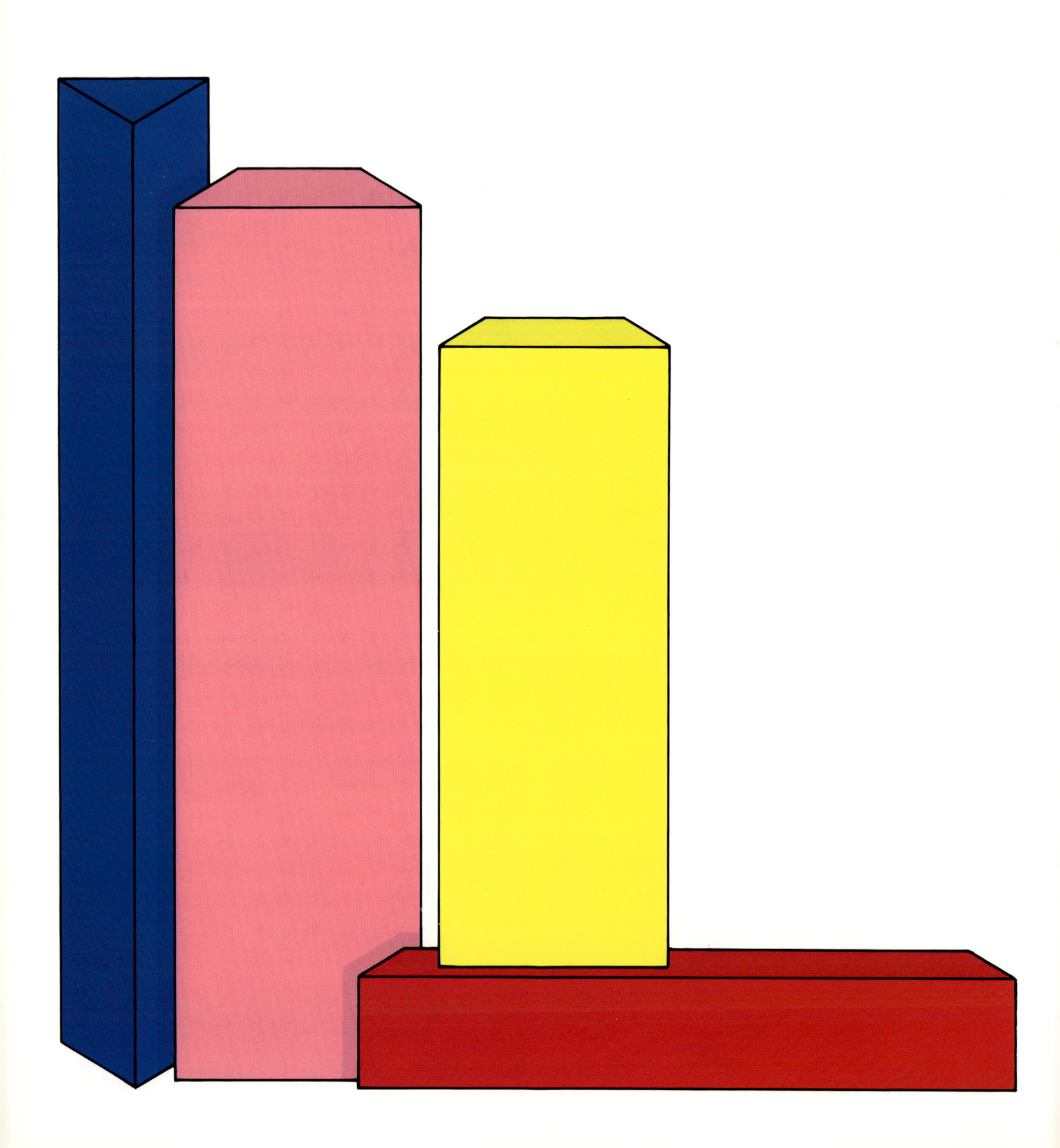

As a first time judge for the PDC Gold awards I was overwhelmed by the quality of submitted work as well as the exceptional organization of the competition by Lea Liu and Charlie Biondo. In many categories the number of exceptional designs created an inability to select only one winner. Packages that stand out for not only their strong graphic appeal but also meeting their stated marketing objectives include: The Cambridge Formula Diet, A.M. Gourmet, Nama-Sake and Ajinomoto Seasonings.

The Cambridge Formula Diet package demonstrates the difference between mass market packaging that requires immediate name identification and direct sales packaging that can concentrate on overall impression. This design's impact would entice even the supermarket consumer to take a closer look and then purchase. The strong luscious closeup fruit photography creates a wonderful color impact and flavor differentiation.

A. M. Gourmet combines both appetizing closeup food photography with product name identification. The large inset photograph against a consistent background unifies the package shelf presence while allowing easy product differentiation.

Nama-Sake and *Ajinomoto* packages demonstrate the uniqueness found in Japanese designs. Simplicity and elegance best describe the Nama-Sake design with frosted etching used as a design element along with standard imprinting. The Ajinomoto design clearly considers the needs of the consumer for single units and organization. The attractiveness of the package allows for table use. Both packages are excellent demonstrations of the fulfillment of marketing needs not just design for the sake of design.

With the continued growth and development of the Package Design Council and superior worldwide packaging, the future looks not only very favorable but very pleasing to the eye.

Beverly Markis
Pillsbury Co.

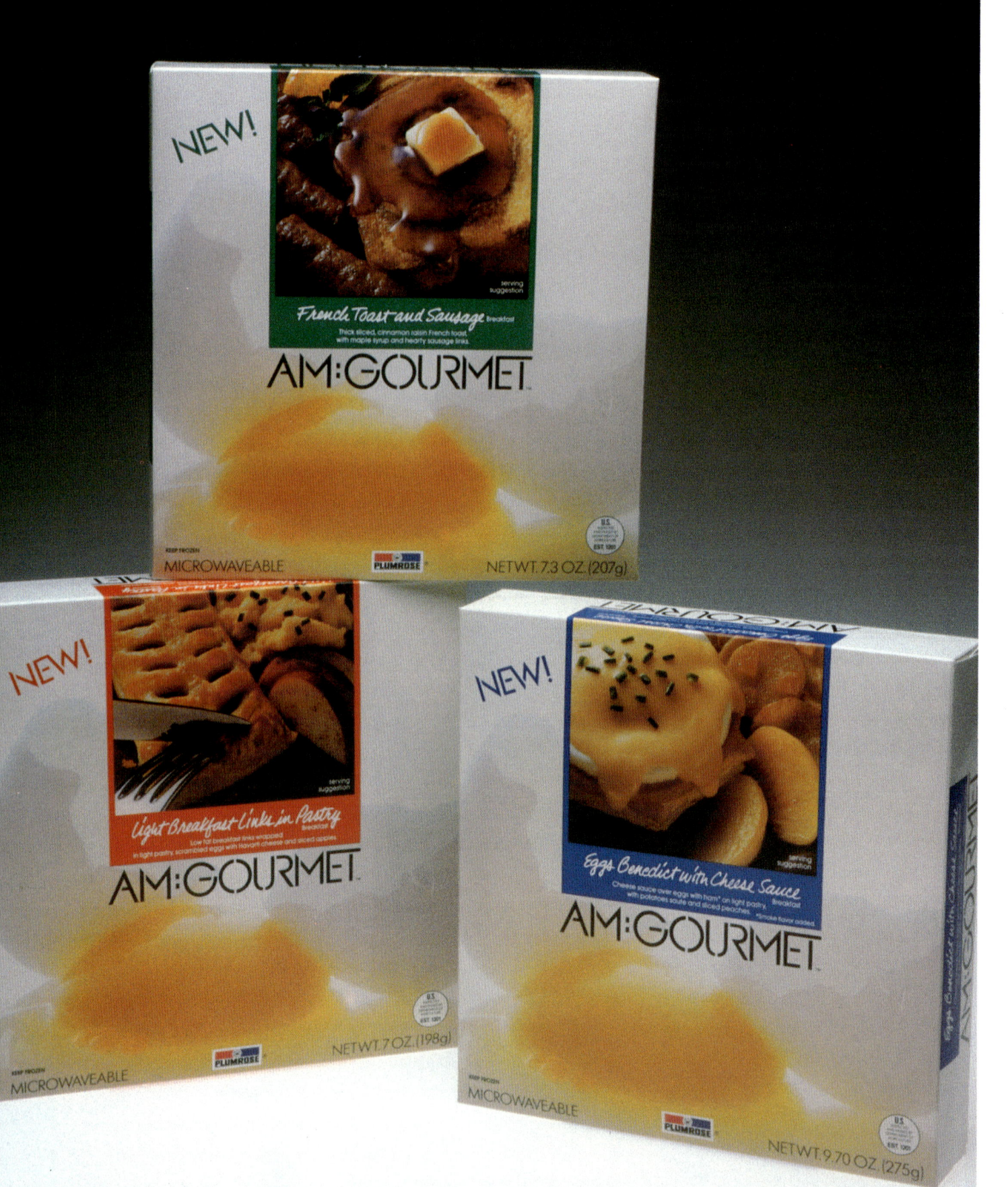

Entry	AM: Gourmet
Category:	Food
Award:	Winner 1987
Division:	USA
Design Firm:	Gerstman+Meyers
Creative Director:	Juan Concepcion
Chief Designer:	Larry Riddell, Sabra Waxman
Design Supervision:	Herbert M. Meyers
Client Firm:	Plumrose

Entry	Panbakes
Category:	Food
Award:	Finalist 1987
Division:	International
Design Firm:	Lewis Moberly
Creative Director:	Mary Lewis
Chief Designer:	Mary Lewis
Client Firm:	Buxted Poultry Ltd.

Entry	Fruit Village In Tosa Jams
Category:	Food
Award:	Finalist 1987
Division:	International
Design Firm:	Design Office Mail Box
Creative Director:	Makoto Umebara
Chief Designer:	Makoto Umebara
Client Firm:	Aoyagi Company

Entry: Grainfield's Cereal
Category: Food
Award: Finalist 1987
Division: USA
Design Firm: Sidjakov Berman Gomez & Partners
Creative Director: Nicolas Sidjakov & Jerry Berman
Chief Designer: James Nevins
Client Firm: Weetabix Company

Entry: Chex Cereal
Category: Food
Award: Finalist 1987
Division: USA
Design Firm: Moonink Communications
Creative Director: John L. Downs
Chief Designer: Moonink Communications
Client Firm: Ralston Purina Company

Entry	Kroger Fresh Pack Preserves
Category:	Food
Award:	Finalist 1987
Division:	USA
Design Firm:	Lipson-Alport-Glass & Associates
Creative Director:	Lipson-Alport-Glass & Associates
Chief Designer:	Lipson-Alport-Glass & Associates
Client Firm:	The Kroger Company

18 Food

Entry:	Weight Watchers Entrees & Desserts
Category:	Food
Award:	Finalist 1987
Division:	USA
Design Firm:	Charles Biondo Design Associates
Creative Director:	Charles Biondo
Client Firm:	Foodways Nationals

Entry	Healthy 'N Light
Category:	Food
Award:	Finalist 1985
Division:	USA
Design Firm:	Landor Associates
Project Director:	Paul Terrill
Creative Director:	Kay Stout
Chief Designer:	Peggy Ng
Client Firm:	Shaklee Corporation

Entry	Le Menu Lightstyle
Category:	Food
Award:	Finalist 1987
Division:	International
Design Firm:	Campbell Soup Company—Design Center
Creative Director:	D. F. Macaulay
Chief Designer:	Barbara Marrington
Client Firm:	Campbell's Frozen Foods Business Unit

20 Food

Entry	La Choy Dinner
Category:	Food
Award:	Finalist 1985
Division:	USA
Design Firm:	Peterson & Blyth Associates Inc.
Creative Director:	David Scarlett
Chief Designer:	David Scarlett
Client Firm:	Beatrice Hunt-Wesson

Entry	Chun King Frozen Dinners
Category:	Foods
Award:	Winner 1980
Design Firm:	Peterson Blyth Associates
Creative Director:	John S. Blyth
Chief Designer:	Peterson & Blyth Design Group
Client Firm:	R. J. Reynolds Foods, Inc.

Food 21

Entry	L'Orient Le Menu
Category:	Food
Award:	Finalist 1987
Division:	International
Design Firm:	Campbell Soup Company—Design Center
Creative Director:	D. F. Macaulay
Chief Designer:	Barbara Marrington
Client Firm:	Campbell's Frozen Foods Business Unit

Entry	Stove Top
Category:	Food
Award:	Finalist 1987
Division:	USA
Design Firm:	Wallace Church Associates
Creative Director:	Stanley Church
Chief Designer:	Phyllis Chan
Design Director:	Marianne McEnrue
Client Firm:	General Foods Corporation

22 Food

Entry	Booth Light & Tender
Category:	Food
Award:	Finalist 1985
Division:	USA
Design Firm:	Lipson-Alport-Glass & Associates (was Lipson Associates at time of award)
Creative Director:	Lipson-Alport-Glass & Associates
Chief Designer:	Lipson-Alport-Glass & Associates
Client Firm:	Booth Seafood Company

Entry	Mrs. Paul's Light Fillets
Category:	Food
Award:	Finalist 1987
Division:	USA
Design Firm:	Lister Butler Inc.
Creative Director:	John Lister
Chief Designer:	Chris Merwin
Client Firm:	Mrs. Paul's Kitchens

Food 23

Entry	Lipton's Potatoes and Sauce
Category:	Food
Award:	Finalist 1987
Division:	USA
Design Firm:	Lister Butler Inc.
Creative Director:	John Lister
Chief Designer:	Catherine Levine
Client Firm:	Thomas J. Lipton

Entry	Mrs. Paul's Seafood En Croute
Category:	Food
Award:	Finalist 1985
Division:	USA
Design Firm:	Dixon & Parcels
Creative Director:	
Chief Designer:	
Client Firm:	Mrs. Paul's Kitchens, Inc.: A Subsidiary of Campbell Soup Company

Entry	Sargento Cheese Line
Category:	Food
Award:	Finalist 1987
Division:	USA
Design Firm:	Murrie, White, Drummond & Lienhart
Chief Designer:	Thomas Q. White
Client Firm:	Sargento Cheese Company, Inc.

Food 25

Entry	Boots Vegetarian Ready Meals
Category:	Food
Award:	Finalist 1987
Division:	International
Design Firm:	Lewis Moberly
Creative Director:	Mary Lewis
Chief Designer:	Mary Lewis
Client Firm:	The Boots Company PLC

26 Food

Entry:	Impromptu
Category:	Food
Award:	Finalist 1987
Division:	USA
Design Firm:	Charles Biondo Design Associates
Creative Director:	Charles Biondo
Client Firm:	General Foods Corp.

Entry	Celentano Entrees
Category:	Food
Award:	Finalist 1987
Division:	USA
Design Firm:	Gerstman+Meyers
Creative Director:	Juan Concepcion
Chief Designer:	Judith Miller
✤✤Design Supervision:	Richard Gerstman
Client Firm:	Celentano Brothers

Entry	Pizzeria
Category:	Food
Award:	Finalist 1985
Division:	USA
Design Firm:	Harte Yamashita & Forest
Creative Director:	Tets Yamashita
Chief Designer:	Susan Healy
Client Firm:	Carnation Company

28 Food

Entry: Ronzoni Product Line
Category: Food
Award: Finalist 1987
Division: USA
Design Firm: Charles Biondo Design Associates
Creative Director: Charles Biondo
Client Firm: General Foods Corp.

Food 29

30 Food

Entry	Panettone Nannini
Category:	Specialty
Award:	Finalist 1987
Division:	International
Design Firm:	Giancarlo Italo Marchi
Creative Director:	Giancarlo Italo Marchi
Client Firm:	Nannini Panettone

Entry	Original San Francisco Sourdough
Category:	Food
Award:	Finalist 1987
Division:	USA
Design Firm:	Sidjakov Berman Gomez & Partners
Creative Director:	Nicolas Sidjakov, Jerry Berman
Chief Designer:	Mark Bergman
Client Firm:	Kraft Foods

Entry	Bread Du Jour
Category:	Food
Award:	Finalist 1987
Division:	USA
Design Firm:	Lister Butler Inc.
Creative Director:	John Lister
Chief Designer:	Catherine Levine
Client Firm:	Continental Baking Company

32 Food

Entry	Wishbone Lite Salad Dressing
Category:	Food
Award:	Finalist 1987
Division:	USA
Design Firm:	Coleman, Lipuma, Segal, & Morrill Inc.
Creative Director:	Owen Coleman
Chief Designer:	Abe Segal
Client Firm:	Thomas J. Lipton Inc.

Entry	Lipton's Cool Side Salads
Category:	Food
Award:	Finalist 1987
Division:	USA
Design Firm:	Lister Butler Inc.
Creative Director:	John Lister
Chief Designer:	Ann Turresson
Client Firm:	Thomas J. Lipton

Entry	Kentucky Fried Chicken Salad Pack
Category:	Miscellaneous
Award:	Finalist 1987
Division:	International
Design Firm:	McCann-Erickson Hakuhodo
Creative Director:	Fumio Kondo
Chief Designer:	Shinji Sudo
Client Firm:	Kentucky Fried Chicken Japan, Ltd.

34 Food

Entry	Campbell's Fresh Tomatoes
Category:	Food
Award:	Finalist 1987
Division:	International
Design Firm:	Campbell Soup Company—Design Center
Creative Director:	D. F. Macaulay
Chief Designer:	William Sterling
Client Firm:	Campbell's Fresh Business Unit

Entry	Green Giant Salads
Category:	Food
Award:	Finalist 1987
Division:	USA
Design Firm:	Nason Design Associates, Inc.
Creative Director:	David Sandstrom/Beverly Markis
Chief Designer:	Brian Rogers
Client Firm:	Pillsbury Center

Entry	Frima Frozen Soup Range
Category:	Food
Award:	Finalist 1987
Division:	International
Design Firm:	Pineapple Design
Creative Director:	Rowland S. W. Heming
Chief Designer:	Isabelle Moreau
Client Firm:	McCain Frima

36 Food

Entry	Ajinomoto Salad Oil
Category:	Specialty
Award:	Winner 1987
Division:	International
Design Firm:	Yao Design Institute Inc.
Creative Director:	Takeo Yao
Client Firm:	Ajinomoto Co., Inc.

Entry:	J Variety of Ajinomoto Gifts
Category:	Specialty
Award:	Finalist 1985
Division:	International
Design Firm:	Ajinomoto Co.

Entry	Monoprix Gourmet
Category:	Food
Award:	Finalist 1987
Division:	International
Design Firm:	Beautiful Design House
Creative Director:	Francis Metzler
Chief Designer:	Francis Metzler, Andre Serrel
Client Firm:	Monoprix

38 Food

Entry	Banshu Somen (Japanese Noodles)
Category:	Specialty
Award:	Finalist 1985
Division:	International
Design Firm:	Par Art Co., Ltd.
Creative Director:	Shiro Tazumi
Chief Designer:	Shiro Tazumi
Client Firm:	Kitamura Co., Ltd.

Entry	Los Angeles 84 Olympics Food Packaging Program
Category:	Specialty
Award:	Finalist 1985
Division:	USA
Design Firm:	Bright & Associates
Creative Director:	Keith Bright/Larry Klein
Chief Designer:	Raymond Wood
Client Firm:	Los Angeles Olympic Organizing Committee (LAOOC)

Food 39

Entry	Namizukushi Cookies
Category:	Food
Award:	Winner 1987
Division:	International
Design Firm:	Yonetsu Design House
Creative Director:	Hisakichi Yonetsu
Chief Designer:	Hisakichi Yonetsu
Client Firm:	Bankaku Sohonpo & Co.

40 Food

Entry	Posada
Category:	Food
Award:	Winner 1985
Division:	USA
Design Firm:	Harte Yamashita & Forest
Creative Director:	Tets Yamashita
Chief Designer:	Janice Barrett
Client Firm:	Centre Brands

Entry	Natural Touch
Category:	Food
Award:	Finalist 1987
Division:	USA
Design Firm:	Libby Perszyk Kathman Inc.
Creative Director:	Ray Perszyk
Chief Designer:	Susan Bailey Zinader
Client Firm:	Worthington Foods

Food 41

Entry:	Casa Fiesta
Category:	Food
Award:	Finalist 1985
Division:	USA
Design Firm:	Sidjakov Berman Gomez & Partners
Creative Director:	Nicolas Sidjakov & Jerry Berman
Chief Designer:	James Nevins
Client Firm:	Bruce Foods Corporation

Entry	Azteca Mexican Foods
Category:	Food
Award:	Finalist 1987
Division:	USA
Design Firm:	Gerstman+Meyers
Creative Director:	Juan Concepcion
Design Supervision:	Richard Gerstman
Graphic Designer:	Alex Pennington
Structural Designer:	Ric Hirst
Client Firm:	The Pillsbury Company

42 Food

Entry	Burger King's Chicken Tenders
Category:	Miscellaneous
Award:	Finalist 1987
Division:	USA
Design Firm:	Gerstman+Meyers
Creative Director:	Juan Concepcion
Chief Designer:	Alex Pennington
Client Firm:	The Burger King Corporation

Entry	Lipton Hearty Cup-A-Soup
Category:	Food
Award:	Finalist 1987
Division:	USA
Design Firm:	Kollberg/Johnson Associates, Inc.
Client Firm:	Thomas J. Lipton

Food 43

Entry	Luzianne Cajun/Creole Foods
Category:	Food
Award:	Finalist 1987
Division:	USA
Design Firm:	Gerstman+Meyers
Creative Director:	Juan Concepcion
Chief Designer:	Karen Corell
❖❖Design Supervision:	Richard Gerstman
Client Firm:	Lizianne Blue Plate Foods

Entry	Knorr Sauce Mix
Category:	Food
Award:	Finalist 1987
Division:	USA
Design Firm:	Irv Koons Associates, Inc.
Creative Director:	Irv Koons
Chief Designer:	Paul Gee
Client Firm:	CPC International

44 Food

Entry	Charlotte Charles Great Impressions
Category:	Food
Award:	Finalist 1987
Division:	USA
Design Firm:	Murrie White Drummond Leinhart & Associates
Chief Designer:	Linda Voll
Client Firm:	Charlotte Charles, Inc.

Entry	A&P Pasta Sauces
Category:	Food
Award:	Finalist 1987
Division:	USA
Design Firm:	The Berni Company
Creative Director:	Stephen H. Berni, PDC
Chief Designer:	Stuart M. Berni, FPDC
Client Firm:	The Great Atlantic & Pacific Tea Company (A&P)

Entry	Heinz Vinegar
Category:	Food
Award:	Finalist 1985
Division:	USA
Design Firm:	Gerstman+Meyers Inc.
Design Supervision:	Richard Gerstman
Creative Director:	Juan Concepcion
Chief Designer:	Joe Lombardo
Client Firm:	H.J. Heinz Company

Entry	Noodles by Leonardo
Category:	Food
Award:	Winner 1981
Design Firm:	Container Corporation of America Design & Market Research Laboratory
Creative Director:	Raymond Peterson
Chief Designer:	Marvin Steck
Client Firm:	Noodles by Leonardo, Inc.

46 Food

Entry	Asakurasansho
Category:	Specialty
Award:	Finalist 1987
Division:	International
Design Firm:	Keizo Matsui And Associates
Creative Director:	Keizo Matsui
Chief Designer:	Sachiko Shinyama
Client Firm:	Fujikko Co., Ltd.

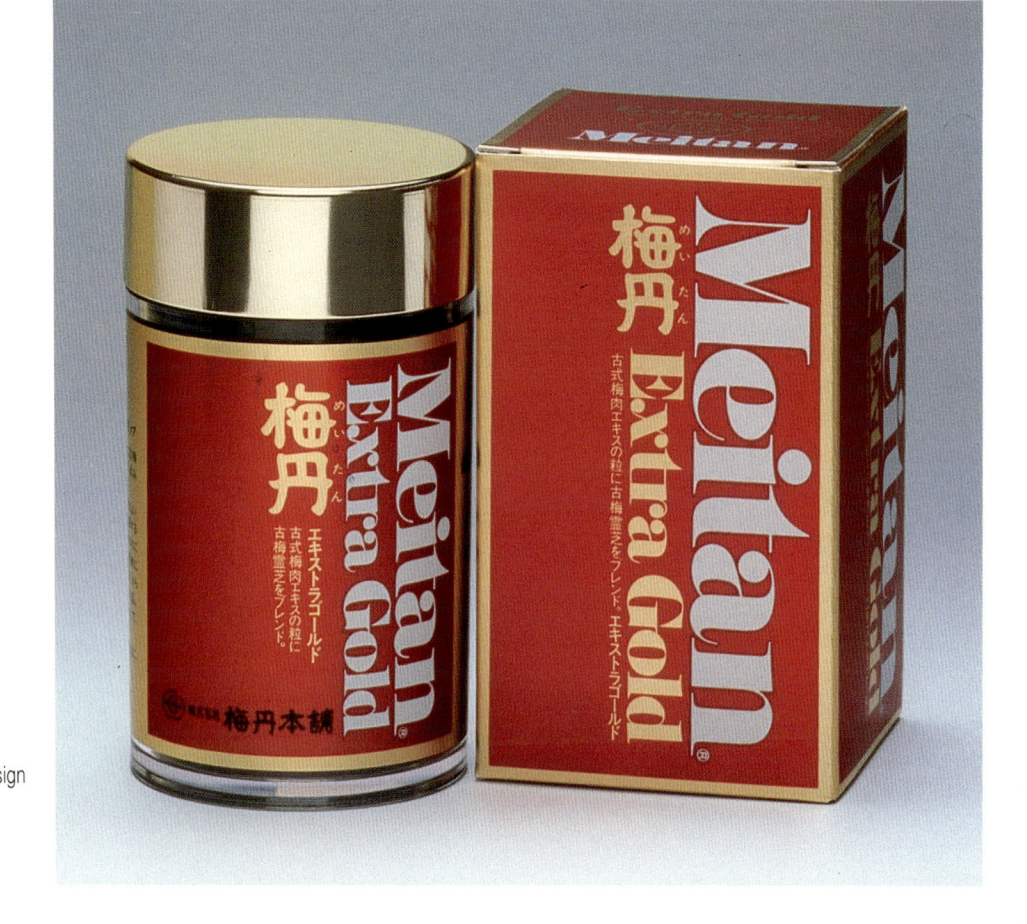

Entry	Meitan Extra Gold
Category:	Food
Award:	Finalist 1985
Division:	International
Design Firm:	Graphic & Package Design Studio SIRO
Creative Director:	Keizo Shiratani
Chief Designer:	Keizo Shiratani
Client Firm:	Meitan Honpo Co., Ltd.

Entry	Ajinomoto Seasoning
Category:	Food
Award:	Winner 1987
Division:	International
Design Firm:	Yao Design Institute Inc.
Creative Director:	Takeo Yao
Chief Designer:	Takuzo Mizuno
Client Firm:	Ajinomoto Co., Inc.

Entry	Celestial Seasonings Spice Blends
Category:	Specialty
Award:	Finalist 1987
Division:	USA
Design Firm:	Lipson-Alport-Glass & Associates
Creative Director:	Lipson-Alport-Glass & Associates
Chief Designer:	Lipson-Alport-Glass & Associates
Client Firm:	Celestial Seasoning/Kraft, Inc.

48 Food

Entry:	Nishikaji 'Goka' (Deluxe) Set
Category:	Specialty
Award:	Finalist 1985
Division:	International
Design Firm:	Rokudo Design Office
Creative Director:	Mitsuru Rokudo
Chief Designer:	Mitsuru Rokudo
Client Firm:	Nishikiaji

Entry	Stella Cheese
Category:	Food
Award:	Finalist 1987
Division:	USA
Design Firm:	Sidjakov Berman Gomez & Partners
Creative Director:	Nicolas Sidjakov, Jerry Berman
Chief Designer:	Dave Curtis
Client Firm:	Universal Foods

Food 49

Entry	Le Doux Brie
Category:	Dairy
Award:	Finalist 1985
Division:	USA
Design Firm:	Libby Perszyk Kathman Inc.
Creative Director:	Jim Gabel
Chief Designer:	Jim Gabel
Client Firm:	The Kroger Co.

50 Food

Entry	Butcher's Blend
Category:	Pet Products
Award:	Winner 1982
Design Firm:	Lister Butler Inc.
Creative Director:	John Lister
Art Director:	Anita Hersh
Chief Designer:	Helen Retger
Client Firm:	Ralston Purina Co.

Entry	Chuck Wagon
Category:	Pet
Award:	Finalist 1985
Division:	USA
Design Firm:	Moonink Communications
Creative Director:	John Downs
Chief Designer:	John Downs
Client Firm:	Ralston Purina Company

Food 51

Entry:	Puppy Kibbles 'N Bits
Category:	Pet
Award:	Finalist 1985
Division:	USA
Design Firm:	Lipson-Alport-Glass & Associates
Creative Director:	Lipson-Alport-Glass & Associates
Chief Designer:	Lipson-Alport-Glass & Associates
Client Firm:	The Quaker Oats Company

Entry	O.N.E.
Category:	Pet Products
Award:	Winner 1987
Division:	USA
Design Firm:	Moonink Communications
Creative Director:	John L. Downs
Chief Designer:	Thomas Jones
Client Firm:	Ralston Purina Company

52 Food

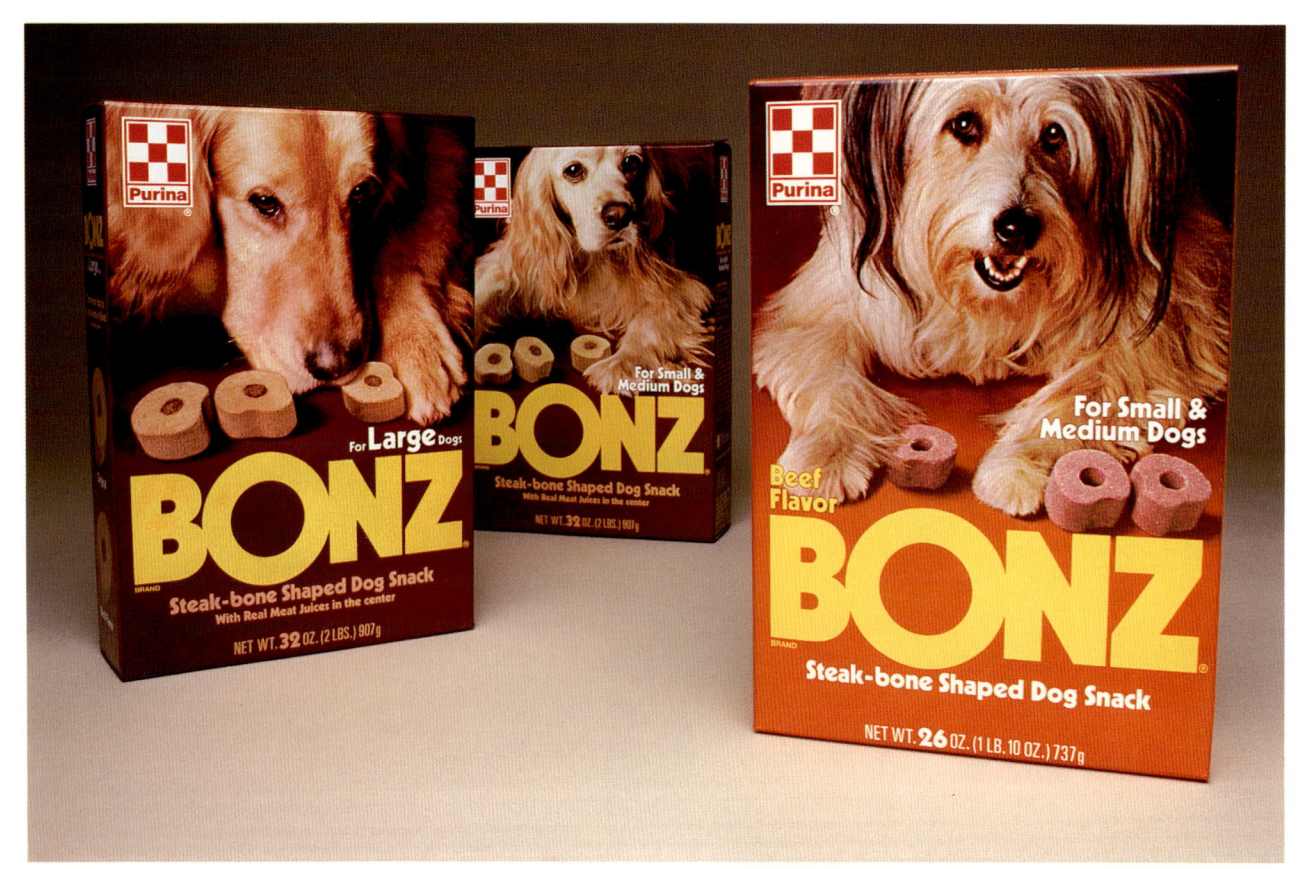

Entry	Bonz
Category:	Pet Products
Award:	Finalist 1985
Division:	USA
Design Firm:	Gerstman+Meyers Inc.
Design Supervision:	Herbert M. Meyers
Creative Director:	Juan Concepcion
Chief Designer:	Rafael Feliciano
Client Firm:	Ralston Purina Company

Entry	Alley Cat
Category:	Pet Products
Award:	Finalist 1985
Division:	USA
Design Firm:	Gerstman+Meyers Inc.
Design Supervision:	Herbert M. Meyers
Creative Director:	Juan Concepcion
Chief Designer:	Ed Szczyglinski
Client Firm:	Ralston Purina Company

Entry	Ralston Purina Horse Food
Category:	Pet Products
Award:	Finalist 1985
Division:	USA
Design Firm:	The Thomas Pigeon Design Group Ltd. (was called Communications Dome at time of award)
Client Firm:	Ralston Purina (Canada) Ltd.

Entry	Kaytee Wild Bird Food
Category:	Pet Products
Award:	Finalist 1987
Division:	USA
Design Firm:	Murrie White Drummond Leinhart & Associates
Chief Designer:	Jeff White
Client Firm:	Kaytee Incorporated

HERSHEY'S
Chocolate & Baked Specialties
CHOCOLATE MEMORIES
Since 1895

Entry	Rich's Dessert Line
Category:	Food
Award:	Winner 1981
Design Firm:	Landor Associates
Client Firm:	Rich Product Corporation

Entry	Hershey's Chocolate Memories
Category:	Specialty
Award:	Finalist 1985
Division:	USA
Design Firm:	Babcock & Schmid Associates, Inc.
Creative Director:	Babcock & Schmid Associates, Inc.
Chief Designer:	Babcock & Schmid Associates, Inc.
Client Firm:	Hershey Food Corporation

56　Food

Entry	Lazzaroni Amaretti Di Saronno
Category:	Specialty
Award:	Finalist 1987
Division:	International
Design Firm:	Campbell Soup Company—Design Center
Creative Director:	D. F. Macaulay
Chief Designer:	John Botthoff
Client Firm:	Campbell's Worldwide Export Co./D. Lazzaroni & C.S.P.A.

Entry	Confectionery Gift
Category:	Specialty
Award:	Finalist 1987
Division:	International
Design Firm:	Par Art Co., Ltd.
Creative Director:	Shiro Tazumi
Chief Designer:	Shiro Tazumi
Client Firm:	Hirakawa Fugetsuda Co., Ltd.

Food 57

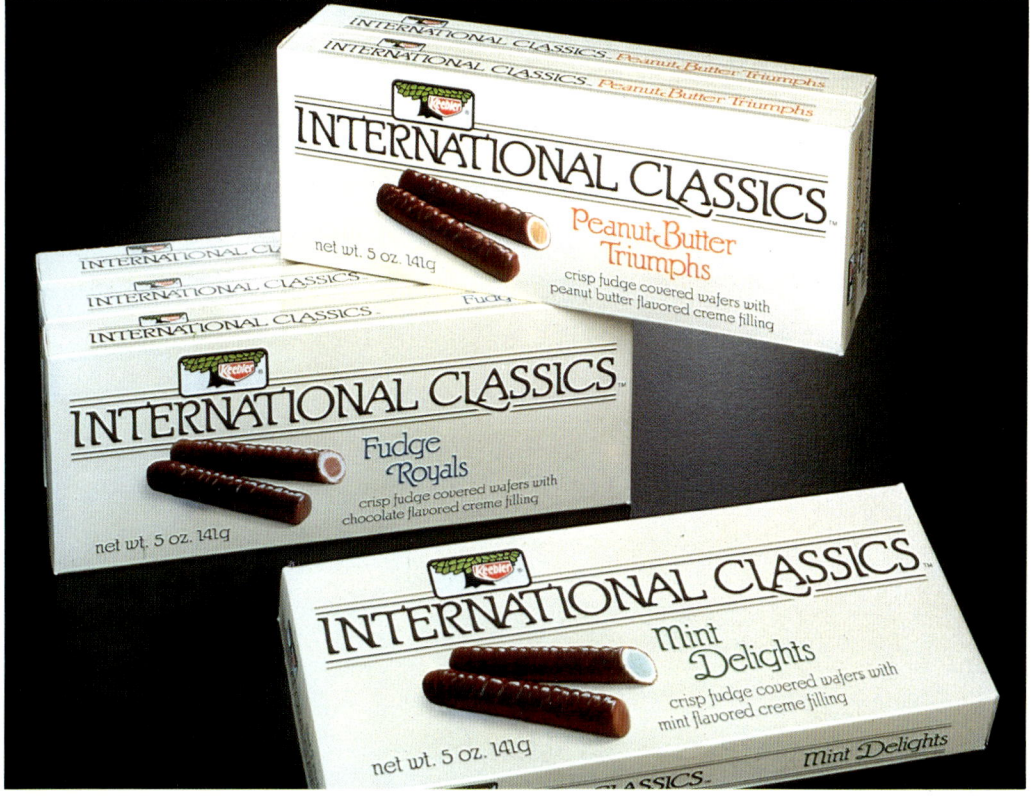

Entry	Morinaga SPUGNA Bitter Milk Chocolate Bar
Category:	Food
Award:	Finalist 1987
Division:	International
Design Firm:	Morinaga & Company Limited Design Room
Creative Director:	Tatsumi Inuzuka
Chief Designer:	Kimiyo Fuji
Client Firm:	Morinaga & Company Limited

Entry:	Keebler International Classics
Category:	Food
Award:	Finalist 1985
Division:	USA
Design Firm:	Wallner, Harbauer, Bruce & Associates
Creative Director:	Jerry Harbauer
Chief Designer:	Jeffery Copp
Client Firm:	The Keebler Company

58 Food

Entry	Metropolitan Opera Gift Candy
Category:	Specialty
Award:	Finalist 1987
Division:	USA
Design Firm:	Dixon & Parcels Associates, Inc.
Creative Director:	J. Roy Parcels
Chief Designer:	Leonora M. Shelsey
Client Firm:	Metropolitan Opera Guild

Entry	LPK Christmas Chocolate Sauce
Category:	Specialty
Award:	Finalist 1987
Division:	USA
Design Firm:	Libby Perszyk Kathman Inc.
Chief Designer:	John Metz
Client Firm:	Libby Perszyk Kathman Inc.

Food 59

Entry	Whitman's Grand Select Chocolates
Category:	Food
Award:	Finalist 1987
Division:	USA
Design Firm:	Gerstman+Meyers
Creative Director:	Juan Concepcion
Chief Designer:	Larry Riddell, Sabra Waxman
Design Supervision:	Herbert M. Meyers
Client Firm:	Whitman's Chocolates

Entry	Michel Guerard Chocolat De Belgique
Category:	Specialty
Award:	Winner 1985
Division:	USA
Design Firm:	The Push Pin Group Inc.
Creative Director:	Seymour Chwast
Chief Designer:	Seymour Chwast, Michael Aron
Client Firm:	Gourmet Resources

60 Food

Entry	Smiths Foods Nut Line
Category:	Food
Award:	Finalist 1985
Division:	International
Design Firm:	Design Board/Behaeghel & Partners S.A.
Creative Director:	Denis Keller
Chief Designer:	Erik Vantal
Client Firm:	Smith Food Group S.A.

Entry	Pop Rocks Candy
Category:	Food
Award:	Finalist 1987
Division:	USA
Design Firm:	Gaylord Adams/Don Flock And Associates Inc.
Creative Director:	Gaylord Adams, Don Flock
Chief Designer:	Gaylord Adams
Client Firm:	Pop Rocks Inc.

Entry	Eagle Snacks Mix
Category:	Food
Award:	Finalist 1987
Division:	USA
Design Firm:	Obata Design, Inc.
Chief Designer:	Dick Lynch
Client Firm:	Eagle Snacks, Inc.

Entry	Henri Nestle Premium Chocolates
Category:	Food
Award:	Winner 1987
Division:	USA
Design Firm:	Lister Butler Inc.
Creative Director:	John Lister
Chief Designer:	Kevin Barnhart
Client Firm:	Nestle Foods Corporation

62 Food

Entry	Bonbon Assorti
Category:	Food
Award:	Finalist 1987
Division:	International
Design Firm:	Yazawa Design Office
Creative Director:	Shozo Yazawa
Chief Designer:	Shozo Yazawa
Client Firm:	Japan Foucher

Entry	Kyo-No-Gozanyaki (Cake)
Category:	Specialty
Award:	Finalist 1985
Division:	International
Design Firm:	Par ARt Co., Ltd.
Creative Director:	Shiro Tazumi
Chief Designer:	Shiro Tazumi
Client Firm:	Goten Yatsuhashi Co., Ltd.

Food 63

Entry	Lazzaroni Amaretti Di Saronno
Category:	Food
Award:	Finalist 1987
Division:	International
Design Firm:	Campbell Soup Company—Design Center
Creative Director:	D. F. Macaulay
Chief Designer:	John Botthoff
Client Firm:	Campbell's Worldwide Export Co./D. Lazzaroni & C.S.P.A.

Entry	KA-FU-HYAKU-SEN
Category:	Food
Award:	Finalist 1987
Division:	International
Design Firm:	GK Graphics Associates
Creative Director:	Shuya Kaneko
Chief Designer:	Eiko Nakayama
Client Firm:	Isetan Department Store, Tokyo

64 Food

Entry:	Riche
Category:	Dairy
Award:	Winner 1982
Design Firm:	Charles Biondo Design Associates
Creative Director:	Charles Biondo
Client Firm:	Venture Foods

Food 65

Entry	Love Bites
Category:	Food
Award:	Finalist 1987
Division:	USA
Design Firm:	Tipton & Maglione
Creative Director:	Judy Tipton
Chief Designer:	Barbara Shelley
Client Firm:	Chipwich

Entry	Yoplait Frozen Yogurt
Category:	Food
Award:	Finalist 1987
Division:	USA
Design Firm:	Peterson & Blyth Associates
Creative Director:	David Scarlett
Chief Designer:	Bob Cruanas
Client Firm:	General Mills

66 Food

Entry	Traditional Biscuit Range
Category:	Food
Award:	Finalist 1987
Division:	International
Design Firm:	Pineapple Design
Creative Director:	Rowland S. W. Heming
Chief Designer:	Rowland S. W. Heming
Client Firm:	Biscuits Delacre

Entry	Grissol Canape Crackers
Category:	Food
Award:	Finalist 1985
Division:	USA
Design Firm:	The Thomas Pigeon Design Group Ltd. (was called Communications Dome at time of award)
Client Firm:	Culinar Foods, Dry Biscuit Division

Food 67

Entry	Rice Cracker Gift
Category:	Specialty
Award:	Finalist 1987
Division:	International
Design Firm:	Par Art Co., Ltd.
Creative Director:	Shiro Tazumi
Chief Designer:	Shiro Tazumi
Client Firm:	Niwa Co., Ltd.

Entry	Clark Bar
Category:	Food
Award:	Finalist 1987
Division:	USA
Design Firm:	Wallner/Harbauer/Bruce & Associates
Creative Director:	Jerry Harbauer
Chief Designer:	Marie V. Grygienc
Client Firm:	Leaf, Inc.

Entry	Asda Private label Chocolates
Category:	Food
Award:	Finalist 1987
Division:	International
Design Firm:	Lloyd Northover Limited
Creative Director:	Simon Ingleton
Chief Designer:	Mark Rollinson
Client Firm:	Asda Stores

Entry	FI-BAR
Category:	Food
Award:	Finalist 1987
Division:	USA
Design Firm:	Harte Yamashita & Forest
Creative Director:	Tets Yamashita
Chief Designer:	Susan Healy
Client Firm:	Natural Nectar Corporation

70 Food

Entry	Majani Candies
Category:	Specialty
Award:	Finalist 1987
Division:	International
Design Firm:	Giancarlo Italo Marchi
Creative Director:	Giancarlo Italo Marchi
Chief Designer:	Giancarlo Italo Marchi
Client Firm:	Majani Candies

Entry	Peanut Butter Krunchies
Category:	Food
Award:	Finalist 1987
Division:	USA
Design Firm:	Lam Design Associates, Inc.
Creative Director:	Michael Lafortezza
Chief Designer:	Lam Design Associates Design Department
Client Firm:	Nestle Foods Corporation

Food 71

Entry	Pillsbury Microwave Popcorn
Category:	Food
Award:	Finalist 1987
Division:	USA
Design Firm:	Wallace Church Associates
Creative Director:	Stanley Church
Chief Designer:	Leslie Tucker
Design Director:	Beverly Markis
Client Firm:	The Pillsbury Company

72 Food

Entry	Soumonka (Confections)
Category:	Specialty
Award:	Winner 1985
Division:	International
Design Firm:	Maeda Design Associates
Creative Director:	Kazuki Maeda
Chief Designer:	Seiichi Maeda
Client Firm:	Souhonke Surugaya

Food 73

Entry	Mifuji Side Dishes
Category:	Miscellaneous
Award:	Finalist 1987
Division:	International
Design Firm:	Keizo Matsui And Associates
Creative Director:	Keizo Matsui
Chief Designer:	Sachiko Shinyama
Client Firm:	Fujikko Co., Ltd.

Entry:	Mousse Chez Nous
Category:	Food
Award:	Winner 1985
Division:	International
Design Firm:	Yao Design Institute Inc.
Creative Director:	Takeo Yao
Chief Designer:	Takuzo Mizuno
Client Firm:	Francais Confectionery Co., Ltd.

74 Food

Entry	Entenmann's Cookies
Category:	Food
Award:	Finalist 1985
Division:	USA
Design Firm:	Peterson & Blyth Associates Inc.
Creative Director:	Peterson & Blyth Associates Inc.
Chief Designer:	Peterson & Blyth Associates Inc.
Client Firm:	Entenmann's

Entry	Perugina Maitre Patissier
Category:	Food
Award:	Finalist 1987
Division:	International
Design Firm:	Giancarlo Italo Marchi
Creative Director:	Giancarlo Italo Marchi
Studio Photographic:	Iris Color
Client Firm:	Perugina Maitre Patissier

Entry:	Whiteday Candy, Cookies, Marshmallows
Category:	Food
Award:	Finalist 1987
Division:	International
Design Firm:	Shincom Co., Ltd.

76 Food

Entry: Edoichi Sembei Confections
Category: Food
Award: Winner 1982
Design Firm: Zero Designs

Entry: Fuwaka Crackers
Category: Specialty
Award: Finalist 1987
Division: International
Creative Director: Hisakichi Yonetsu
Chief Designer: Hisakichi Yonetsu
Client Firm: Bankaku Sohonpo & Co.

Food 77

Entry:	Prestige Lideseal
Category:	Dairy
Award:	Finalist 1985
Division:	USA
Design Firm:	James River Corporation

Entry	Breyers Ice Cream
Category:	Dairy
Award:	Winner 1985
Division:	USA
Design Firm:	Gerstman+Meyers Inc.
Design Supervision:	Richard Gerstman
Creative Director:	Juan Concepcion
Chief Designer:	Larry Riddell
Client Firm:	Kraft, Inc. Dairy Group

80 Food

Entry: Nestle Premium Ice Cream
Category: Food A
Award: Finalist 1987
Division: USA
Design Firm: Nestle Foods Corp.

Entry	DoveBars Two-pack
Category:	Food
Award:	Finalist 1987
Division:	USA
Design Firm:	Lipson-Alport-Glass & Associates
Creative Director:	Lipson-Alport-Glass & Associates
Chief Designer:	Lipson-Alport-Glass & Associates
Client Firm:	Dove International/Division of Mars, Inc.

Entry	Forever Yours
Category:	Food
Award:	Finalist 1987
Division:	USA
Design Firm:	Lipson-Alport-Glass & Associates
Creative Director:	Lipson-Alport-Glass & Associates
Chief Designer:	Lipson-Alport-Glass & Associates
Client Firm:	Dove International/Division of Mars, Inc.

82 Food

Entry	Arctic Fruit
Category:	Food
Award:	Finalist 1987
Division:	USA
Design Firm:	Source/Inc.
Creative Director:	Guy Gangi
Chief Designer:	Hedy Lim Wong, Kim Read Vallone
Client Firm:	The Quaker Oats Co.

Entry	Chiquita Pops
Category:	Food
Award:	Finalist 1985
Division:	USA
Design Firm:	The Berni Company
Creative Director:	Stephen H. Berni, PDC
Chief Designer:	Stuart M. Berni, FPDC
Client Firm:	United Brands Company

Food 83

Entry	Rondos Ice Cream Chocolates
Category:	Food
Award:	Finalist 1987
Division:	USA
Design Firm:	Lipson-Alport-Glass & Associates
Creative Director:	Lipson-Alport-Glass & Associates
Chief Designer:	Lipson-Alport-Glass & Associates
Client Firm:	Dove International/Division of Mars, Inc.

84　Food

Entry:	Dole Sorbet
Category:	Dairy
Award:	Finalist 1985
Division:	USA
Design Firm:	Charles Biondo Associates
Creative Director:	Charles Biondo
Chief Designer:	
Client Firm:	Dole Foods

Food 85

Entry	Fruit Spoonables
Category:	Food
Award:	Finalist 1987
Division:	USA
Design Firm:	Kollberg/Johnson Associates, Inc.
Client Firm:	Ocean Spray Cranberries

CHAPTER 2

Beverage

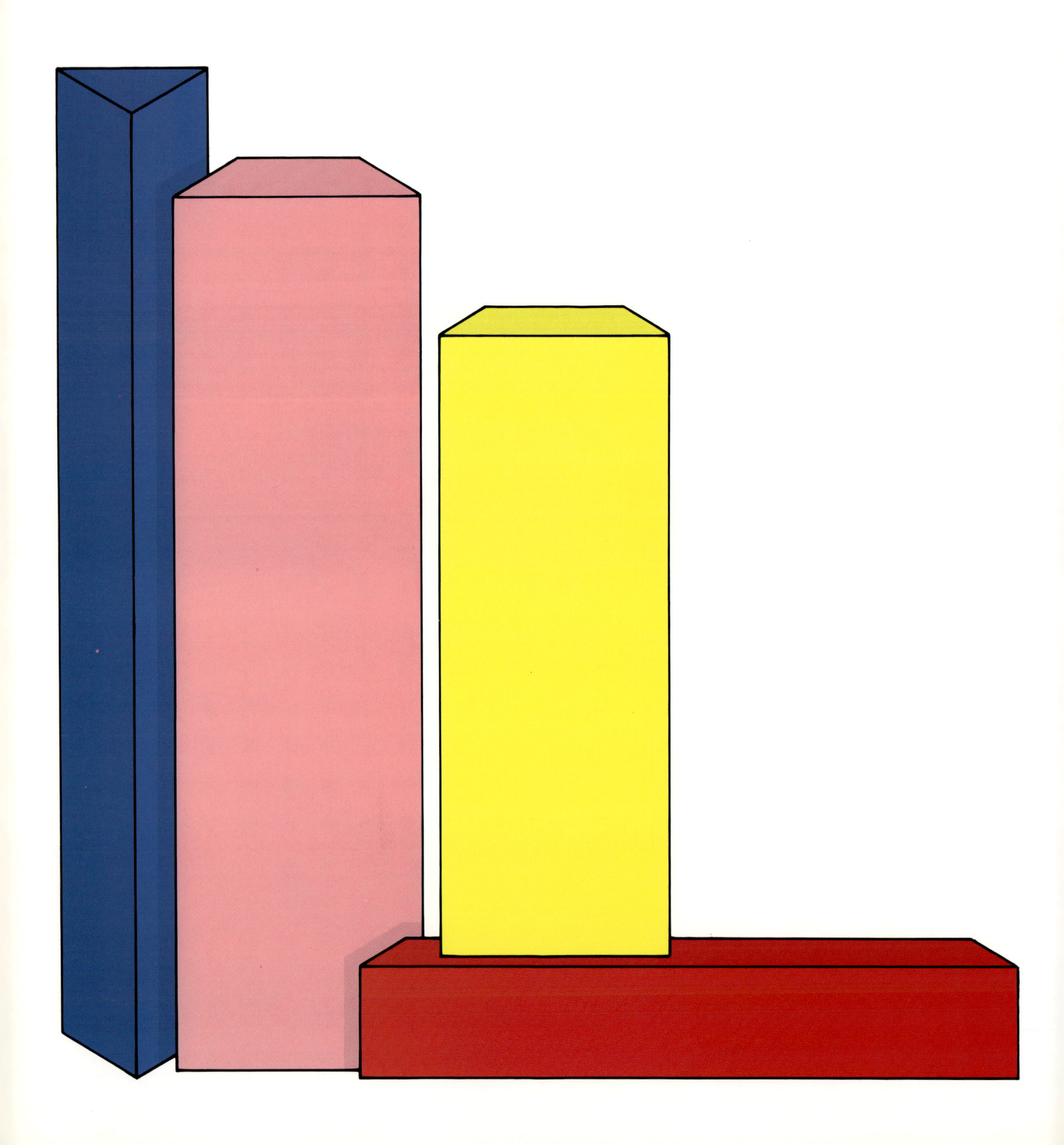

Comments on the 1985 and 1987 PDC Gold Award Competitions

When judging the *1985 Gold Award Competition*, I was impressed by the sheer number of entries and the diversity of the graphic techniques portrayed on the packages. Several trends also became apparent. The Helvetica typeface, so popular on packages a few years ago, had become practically non-existent. Logos and lettering styles were striving for their own character on most of the entries. Black backgrounds were no longer "taboo" on packages, and these backgrounds allowed other package elements to achieve strong color impact.

The 1985 winners had a commonality in their architectural graphic structure. The graphic elements on the winners were minimal, strong, clear and symmetrical. Perhaps Gold Award competition judges think alike, or 1985 tastes manifested themselves on packaging in a similar fashion. Or maybe marketing conditions of 1985 demanded certain kinds of packaging with similar traits.

I always judge packages along certain criteria: overall communication regarding the product inside, imagery, layout and most importantly—the "big idea." A good example of ideas that strongly communicate were displayed by two winners—Breyers Ice Cream and Clear Elegance packages. On each of these packages, the product inside takes on a new dimension. The *Breyers Ice Cream* product photo is "larger than life," scooped for appetite appeal, and displays a garnish that relates to the flavor. There is absolutely no color on the package except for the ice cream and the Breyers leaf.

The *Clear Elegance* dishware package gives the product a new dimension by incorporating its usage in a unique manner, namely with a very colorful and eye-catching lobster.

These consumer-oriented packages functionally identify brand and product, and target their consumers immediately with their end benefit, using "the big idea."

The 1987 Awards were also impressive due to the massive scope of entries in many categories that reflected every possible style of package configuration, graphics, materials and printing methods.

Overall, there was again a creative use of typography that separated this year's packaging from that of several years ago. Designers once again strove to give each its own character, and unique typefaces again became an important part of distinguishing the imagery of one brand from another.

The *1987 International packages* that were most impressive were not impactful, but actually were quite subtle in overall appearance. They relied on very delicate graphic elements to convey their imagery. I responded favorably to the quiet elegance of packages such as the Nama-Sake liquor (wine). This particular design had a unique idea conveyed through its combination of clear and frosted glass, making a bottle-texture pattern that was unusual and totally "unexpected." Color touches were added through the neckband and calligraphy, and one could actually feel the mood of "a quiet evening in Japan."

The 1987 *USA Division* winners comprised more impactful packaging than the International winners. However, the overall mood and theme of the 1987 packages was somewhat restrained compared to previous years. The winning packages reflect a symmetry in graphic architecture, similar to the 1985 winners.

One package group that seemed to separate itself from most of the other entries was *Cambridge Formula Diet* (dairy and dessert foods). Oversized photographs of fruit lined up to form a unique pattern, attract immediate attention and, to me, conveyed naturalness and orderliness—just the imagery that should be conveyed for a proper diet.

I was also impressed by the photograph and the imagery utilized on the *Berkley Sporting Goods* products and the *AM:Gourmet* microwave products. Through clever layout and photography, products that are not themselves exceptional, are, through packaging, able to attract strong attention and immediately convey the end benefit to the consumer.

To make one brand perform better than others in the retail environment, the designer must be able to find that point-of-difference and communicate it on the outside of the package. This instant communication together with a unique idea to attract attention at retail, give a championship quality to packaging. Overall, I believe many of the USA Division winners achieved this.

Richard Gerstman, PDC
Gerstman + Meyers, Inc.

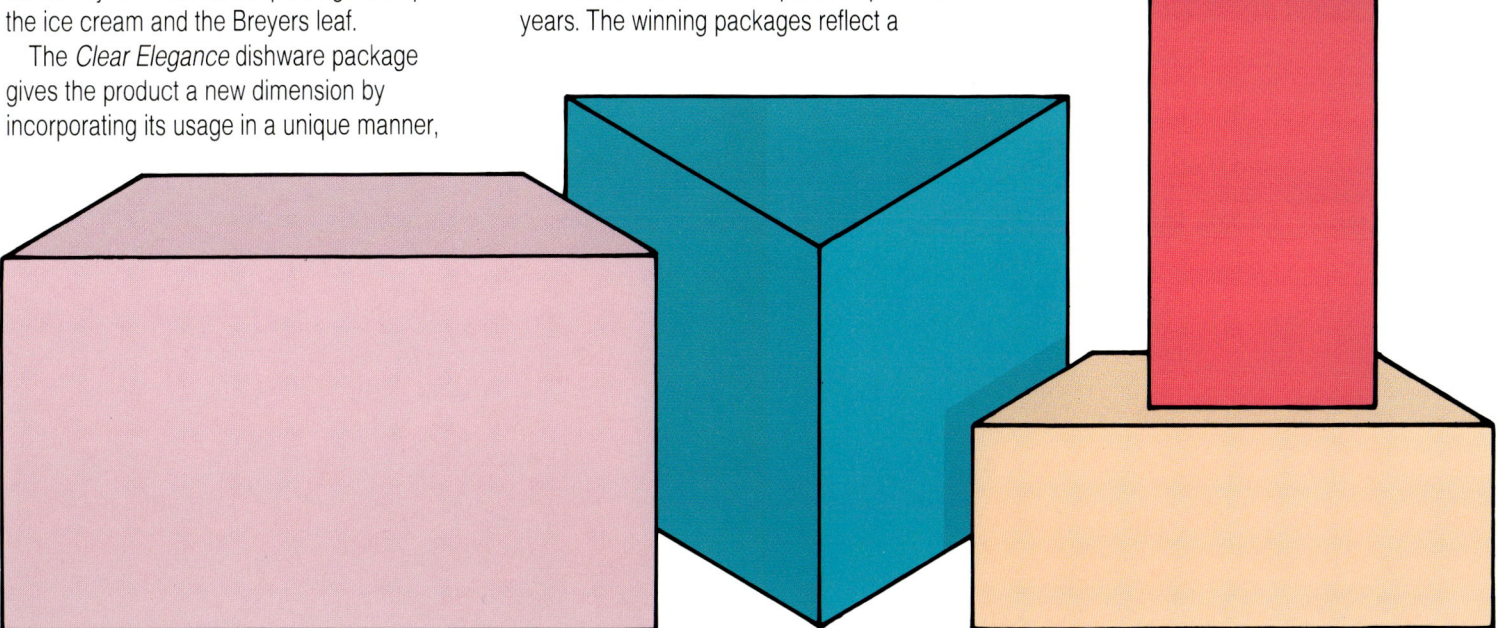

Entry:	Christian Brothers Wine
Category:	Beverage
Award:	Finalist 1985
Division:	USA
Design Firm:	Primo Angeli, Inc.
Creative Director:	Primo Angeli
Chief Designer:	Ray Honda, Eric Read, Mark Jones, Primo Angeli
Client Firm:	The Christian Brothers

Beverage 89

Entry	Chateau Chevalier
Category:	Beverage
Award:	Finalist 1987
Division:	USA
Design Firm:	Colonna, Farrell: Design Associates
Creative Director:	Ralph Colonna
Chief Designer:	Ralph Colonna
Client Firm:	Chateau Chevalier Winery

Entry	Rutherford Estate
Category:	Beverage
Award:	Finalist 1987
Division:	USA
Design Firm:	Colonna, Farrell: Design Associates
Creative Director:	John Farrell
Chief Designer:	Cynthia Maguire
Client Firm:	Rutherford Estate Winery/ Inglenook-Napa Valley

Beverage

Entry	Yumemurasaki
Category:	Specialty
Award:	Finalist 1987
Division:	International
Design Firm:	Rich Co., Ltd.
Creative Director:	Yoshinori Shiozawa
Chief Designer:	Yoshinori Shiozawa
Client Firm:	Tsuji Hoten Co., Ltd.

Entry	Daiginjyo-Touji-No-Sake
Category:	Specialty
Award:	Finalist 1987
Division:	International
Design Firm:	Rich Co., Ltd.
Creative Director:	Makoto Sato
Chief Designer:	Makoto Sato
Client Firm:	Madonoume Shuzo Co., Ltd.

Beverage 91

Entry	Suiginko
Category:	Specialty
Award:	Finalist 1987
Division:	International
Design Firm:	Rich Co., Ltd.
Creative Director:	Makoto Sato
Chief Designer:	Makoto Sato
Client Firm:	Shibata Partnership

Entry	Nama-Sake
Category:	Beverage
Award:	Winner 1987
Division:	International
Design Firm:	Rich Co., Ltd.
Creative Director:	Yoshinori Shiozawa
Chief Designer:	Yoshinori Shiozawa
Client Firm:	Tsuji Hoten Co., Ltd.

92 Beverage

Entry	La Mancha
Category:	Beverage A
Award:	Finalist 1987
Division:	International
Design Firm:	Lewis Moberly
Creative Director:	Mary Lewis
Chief Designer:	Mary Lewis
Client Firm:	ASDA Stores Ltd.

Entry	Cinzano Asti Spumante
Category:	Beverage
Award:	Finalist 1987
Division:	USA
Design Firm:	Kass Communications
Creative Director:	Warren A. Kass
Chief Designer:	Warren A. Kass
Client Firm:	The Paddington Corporation

Beverage 93

94 Beverage

Entry	Gloria Ferrer
Category:	Beverage
Award:	Finalist 1987
Division:	USA
Design Firm:	Colonna, Farrell: Design Associates
Creative Director:	Ralph Colonna
Chief Designer:	Michelle Collier
Client Firm:	Gloria Ferrer Sonoma Champagne Caves

Entry	Cribari Easy-Pour Winepack
Category:	Beverage
Award:	Finalist 1987
Division:	USA
Design Firm:	Jon Wells Graphic Design (carton graphics) Toppan Printing Company (structural design)
Creative Director:	Michael Friedman, Guild Wineries
Chief Designer:	Jon Wells
Client Firm:	Guild Wineries

Entry	Zaca Mesa
Category:	Beverage
Award:	Winner 1987
Division:	USA
Design Firm:	Colonna, Farrell: Design Associates
Creative Director:	Ralph Colonna
Chief Designer:	Amy Racina
Client Firm:	Zaca Mesa Winery

96 Beverage

Entry	Pure Malt Whisky
Category:	Beverage
Award:	Finalist 1987
Division:	International
Design Firm:	Taku Satoh Design Office
Creative Director:	Dentsu Inc.
Chief Designer:	Taku Satoh
Client Firm:	Nikka Whisky Distilling Co., Ltd.

Entry	Jamaica White Rum
Category:	Beverage
Award:	Finalist 1987
Division:	International
Design Firm:	Lewis Moberly
Creative Director:	Mary Lewis
Chief Designer:	Neil Wood
Client Firm:	ASDA Stores Ltd.

Entry	Alfred Lambs Special Reserve
Category:	Beverage
Award:	Finalist 1987
Division:	International
Design Firm:	Lewis Moberly
Creative Director:	Mary Lewis
Chief Designer:	Mary Lewis
Client Firm:	United Rum Merchants (UK) Ltd.

98 Beverage

Entry	Blanton's Kentucky Bourbon
Category:	Beverage
Award:	Finalist 1985
Division:	USA
Design Firm:	The Creative Source, Inc.
Creative Director:	Thomas D'Addario
Chief Designer:	Adam D'Addario, Bill Riedell
Client Firm:	Age International Inc.

Entry	J&B Rare Scotch
Category:	Beverage
Award:	Finalist 1987
Division:	USA
Design Firm:	Werbin Associates
Creative Director:	Irv Werbin
Chief Designer:	Irv Werbin
Client Firm:	The Paddington Corporation

Beverage 99

Entry	Takara Nihon-Kazan
Category:	Beverage
Award:	Finalist 1987
Division:	International
Design Firm:	Graphic & Package Design Studio Siro
Creative Director:	Keizo Shiratani
Chief Designer:	Keizo Shiratani
Client Firm:	Takara Shuzo Co., Ltd.

Entry	Daiginjyo-Miyozakura
Category:	Specialty
Award:	Finalist 1987
Division:	International
Design Firm:	Rich Co., Ltd.
Creative Director:	Yoshinori Shiozawa
Chief Designer:	Yoshinori Shiozawa
Client Firm:	Miyozakura Jyozo Co., Ltd.

Beverage

Beverage 101

Entry Sambuca Molinari
Category: Beverage
Award: Finalist 1987
Division: USA
Design Firm: Mittleman/Robinson Design Associates
Creative Director: Fred Mittleman
Chief Designer: Fred Mittleman
Client Firm: Buckingham Wile

Entry: Welch's Orchard
Category: Beverage B
Award: Winner 1985
Division: USA
Design Firm: Sidjakov Berman Gomez & Partners
Creative Director: Nicolas Sidjakov & Jerry Berman
Chief Designer: James Nevins
Client Firm: Welch Foods, Inc.

Entry Tropic Freezer
Category: Beverage
Award: Finalist 1987
Division: USA
Design Firm: Lam Design Associates, Inc.
Creative Director: Michael Lafortezza
Chief Designer: Lam Design Associates Design Department
Client Firm: Heublein

102 Beverage

Entry	Treesweet Lite
Category:	Beverage
Award:	Finalist 1987
Division:	USA
Design Firm:	Primo Angeli Inc.
Creative Director:	Primo Angeli
Chief Designer:	Primo Angeli, Ray Honda
Client Firm:	Treesweet Products Co.

Entry	Sunflo Juice
Category:	Beverage
Award:	Finalist 1987
Division:	USA
Design Firm:	Selame Design
Creative Director:	Dave Adams
Chief Designer:	Joseph Selame
Client Firm:	Sunflo Juice

Beverage 103

Entry	Sunkist Orange Juice
Category:	Non-Alcoholic Beverage
Award:	Finalist 1985
Division:	USA
Design Firm:	Gerstman+Meyers Inc.
Design Supervision:	Richard Gerstman
Creative Director:	Juan Concepcion
Chief Designer:	Larry Riddell
Client Firm:	Thomas J. Lipton, Inc.

Entry	Chiquita Orange Banana Juice
Category:	Beverage
Award:	Finalist 1987
Division:	USA
Design Firm:	The Berni Company
Creative Director:	Stuart M. Berni, FPDC
Chief Designer:	Mark Eckstein
Client Firm:	United Brands Company

104 Beverage

Entry	Treesweet
Category:	Beverage
Award:	Finalist 1987
Division:	USA
Design Firm:	Primo Angeli Inc.
Creative Director:	Primo Angeli
Chief Designer:	Primo Angeli, Vicki Cero, Ray Honda, Mark Jones
Client Firm:	Treesweet Products Co.

Entry	Lipton Fruit Tea
Category:	Beverage
Award:	Finalist 1987
Division:	USA
Design Firm:	Kollberg/Johnson Associates, Inc.
Client Firm:	Thomas J. Lipton

Beverage 105

Entry	Minute Maid Sirop Range
Category:	Beverage
Award:	Finalist 1987
Division:	International
Design Firm:	Pineapple Design
Creative Director:	Rowland S. W. Heming
Chief Designer:	Daniele Matthys
Client Firm:	Beverage Products Ltd.

Entry	Refresh
Category:	Beverage
Award:	Finalist 1987
Division:	USA
Design Firm:	Source/Inc.
Creative Director:	Charles Cybul
Chief Designer:	Bernard Dolph
Client Firm:	The Quaker Oats Co.

106 Beverage

Entry	Lincoln Juices
Category:	Beverage
Award:	Finalist 1987
Division:	USA
Design Firm:	The Berni Company
Creative Director:	Stuart M. Berni, FPDC
Chief Designer:	Jung Kim
Client Firm:	Sundor Brands, Inc.

Entry	Nestle Hot Cocoa
Category:	Beverage
Award:	Finalist 1987
Division:	USA
Design Firm:	Joel Bronz Design Inc.
Creative Director:	Joel Bronz
Chief Designer:	Karen Willoughby
Client Firm:	Nestle Foods Corporation

Beverage 107

Entry	Vital 15
Category:	Beverage
Award:	Finalist 1987
Division:	USA
Design Firm:	Sidjakov Berman Gomez & Partners
Creative Director:	Nicolas Sidjakov, Jerry Berman
Chief Designer:	Barbara Vick
Client Firm:	California Milk Advisory Board

Entry:	Famous Amos Rich 'N Famous Hot Cocoa Mix
Category:	Beverage
Award:	Finalist 1987
Client Firm:	Van Dutch

108 Beverage

Entry:	La Vie The (Tea)
Category:	Beverages
Award:	Finalist 1987
Division:	International
Design Firm:	C'Bon Cosmetics Co., Ltd.

Entry	KYO-NO OCHADOKORO Japanese Tea
Category:	Non-Alcoholic Beverages
Award:	Winner 1982
Design Firm:	Par Art Room

Beverage 109

Entry	Sanka
Category:	Beverage
Award:	Finalist 1987
Division:	USA
Design Firm:	Wallace Church Associates
Creative Director:	Robert Wallace
Chief Designer:	Phyllis Chan
Design Director:	Joella Daunys
Client Firm:	General Foods Corporation

Entry	Blendy Water Soluble Coffee
Category:	Beverage
Award:	Finalist 1987
Division:	International
Design Firm:	Shin Matsunaga Design Office Inc. & Motomi Kawakami Design Room
Creative Director:	Hajime Tajima
Chief Designer:	Shin Matsunaga, Motomi Kawakami
Client Firm:	Ajinomoto General Foods, Inc.

110 Beverage

Entry	Hills Brothers Flavor Roast
Category:	Beverage
Award:	Finalist 1987
Division:	USA
Design Firm:	Sidjakov Berman Gomez & Partners
Creative Director:	Nicolas Sidjakov, Jerry Berman
Chief Designer:	Jackie Foshuag
Client Firm:	Hills Brothers Coffee

Entry	Chase & Sanborn Instant Coffee
Category:	Beverage
Award:	Finalist 1987
Division:	USA
Design Firm:	Sidjakov Berman Gomez & Partners
Creative Director:	Nicolas Sidjakov, Jerry Berman
Chief Designer:	Thomas Bond
Client Firm:	Hills Brothers Coffee

Entry	Night & Day
Category:	Beverage
Award:	Winner 1987
Division:	International
Design Firm:	Lewis Moberly
Creative Director:	Mary Lewis
Chief Designer:	Jerome Berard
Client Firm:	Foods Brands Groups Ltd.

112 Beverage

Entry	The Cellar Coffee Products
Category:	Miscellaneous
Award:	Finalist 1987
Division:	USA
Design Firm:	RH Macy & Co., Inc. Corporate Graphics
Creative Director:	Marion Fraternale
Chief Designer:	Robert Tabor
Client Firm:	RH Macy & Co., Inc.

Entry	Johann Jacobs Coffee
Category:	Beverage
Award:	Finalist 1987
Division:	USA
Design Firm:	S&O Consultants Inc.
Creative Director:	Edwin Ewry
Chief Designer:	Diane Nelson
Client Firm:	The Jacobs Coffee Company

Entry:	Chase & Sanborn
Category:	Beverage B
Award:	Finalist 1985
Division:	USA
Design Firm:	Sidjakov Berman Gomez & Partners
Creative Director:	Nicolas Sidjakov & Jerry Berman
Chief Designer:	James Nevins
Client Firm:	Chase & Sanborn Foods

Entry	Maxwell House Private Collection
Category:	Beverage
Award:	Winner 1987
Division:	USA
Design Firm:	General Foods Corporate Design Center
Creative Director:	Louise Calo
Chief Designer:	Peter Nikolits
Client Firm:	Maxwell House Division of General Foods Corporation

114 Beverage

Entry	Montclair Sparkling Mineral Water
Category:	Non Alcoholic Beverage
Award:	Winner 1980
Design Firm:	Gerstman+Meyers
Design Supervision:	Herbert M. Meyers
Creative Director:	Juan Concepcion
Chief Designer:	Gerstman+Meyers design staff
Client Firm:	Allan Beverage Co. Ltd.

Entry	Hawaiian Water
Category:	Beverage
Award:	Finalist 1987
Division:	USA
Design Firm:	Landor Associates
Creative Director:	Kay Stout
Chief Designer:	Lotis Cajoleas
Client Firm:	Pure Ltd./Hawaiian Water Partners

Entry	Fletcher & Oakes Schnapps
Category:	Beverage
Award:	Finalist 1987
Division:	USA
Design Firm:	Source/Inc.
Creative Director:	Charles Cybul
Chief Designer:	Dale Joneson
Client Firm:	James B. Beam Distilling Co.

116 Beverage

Entry:	Erlanger Beer
Category:	Beverage A
Award:	Winner 1985
Division:	USA
Design Firm:	The Pushpin Group Inc.
Creative Director:	Seymour Chwast
Chief Designer:	Seymour Chwast, Michael Aron
Client Firm:	MCA Advertising for Stroh Brewery

Entry	Winterfest
Category:	Beverage
Award:	Finalist 1987
Division:	USA
Design Firm:	Libby Perszyk Kathman Inc.
Creative Director:	Ray Perszyk, Jeff Copp
Chief Designer:	John Metz
Client Firm:	Adolph Coors Company

Entry	Henry Weinhard Light
Category:	Beverage
Award:	Finalist 1987
Division:	USA
Design Firm:	Primo Angeli Inc.
Creative Director:	Primo Angeli
Chief Designer:	Primo Angeli, Mark Jones
Client Firm:	Hal Riney & Partners (for Blitz Weinhard Brewing Co.)

118 Beverage

Entry	Anheuser Beer
Category:	Beverage
Award:	Finalist 1987
Division:	USA
Design Firm:	Obata Design, Inc.
Chief Designer:	Dick Lynch
Client Firm:	Anheuser-Busch, Inc.

Beverage 119

Entry	Takara Barbican
Category:	Beverage
Award:	Finalist 1987
Division:	International
Design Firm:	Graphic & Package Design Studio Siro
Creative Director:	Keizo Shiratani
Chief Designer:	Keizo Shiratani
Client Firm:	Takara Shuzo Co., Ltd.

120 Beverage

Entry:	Herman Joseph's 1868
Category:	Beverages
Award:	Winner 1981
Design Firm:	Demartin Marona Cranstoun Downes
Creative Director:	Robert Marona
Chief Designer:	Ed Demartin
Client Firm:	Adolph Coors Company

Entry	Miller Beer Holiday Packaging
Category:	Specialty
Award:	Finalist 1985
Division:	USA
Design Firm:	Design North
Creative Director:	Berni Spinelli, Miller Brewing Co.
Chief Designer:	Mark Akgulian
Client Firm:	Miller Brewing Company

Entry	Biere Haute Fermentation
Category:	Beverage
Award:	Finalist 1987
Division:	International
Design Firm:	Beautiful Design House
Creative Director:	Olivier Desdoigts
Chief Designer:	Guillaume Toutain
Client Firm:	Brasserie de Solenes

CHAPTER **3**

Health and Beauty Aids

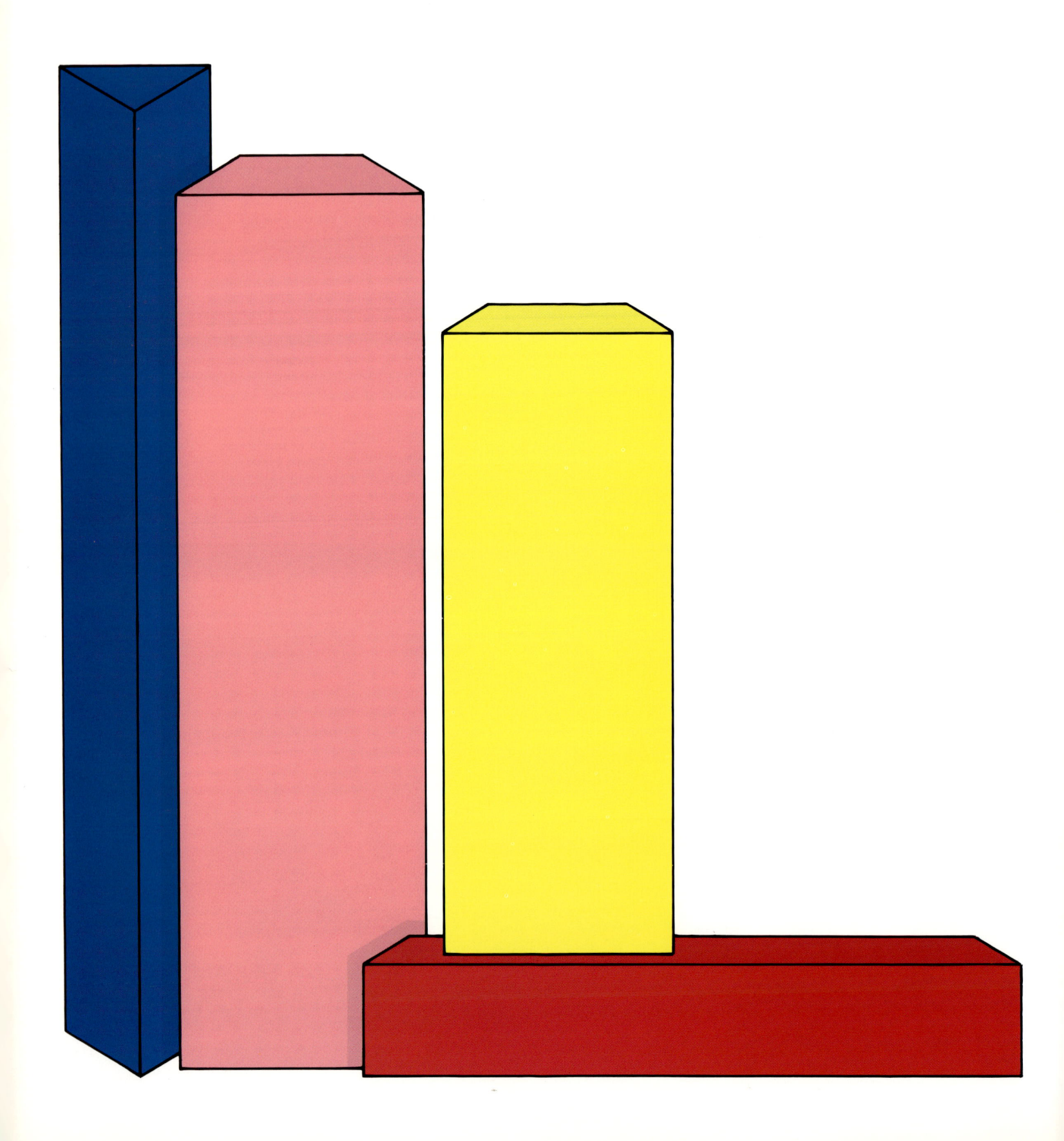

Presently, when individual value judgment is taking an increasingly serious view, mass consumption is rapidly declining in Japan and special brand goods may occupy 30-35% of the entire market.

Under these circumstances, the packaging business should be reexamined, by not just looking at the charm of a package design, but also at the region of motivation behind it. In the case of package production, business changes from visual creation to invisible creation, that is composite creation inclusive of tactual creation. Packaging is not the mere act of packing, but is "goods" itself and an important means of "worth creation."

Highly appreciated Japanese packages in foreign countries are generally traditional in design, but the PDC Gold Award winners recently chosen are extremely internationalized ones. The "design" boundaries placed on Japanese packages have been removed. I believe increasing international exchanges through the PDC Gold Award Competition is useful for refining a broad international sense of the visual field allowing for the existence of various value judgments.

I am very pleased to join the 1987 PDC Gold Award Competition as a judge and to find over a hundred entries from Japan. I was impressed with it as a person who hopes for the establishment of the packaging design business. PDC is the only group in the world organized by packaging design professionals and flourishing activities, including the Gold Award Competition, that attract public attention.

Takeo Yao
Yao Design Institute
PDC Japan Chapter

124   Health and Beauty Aids

Entry: Halsa Shampoo
Category: Health and Beauty Aids
Award: Finalist 1985
Division: USA
Design Firm: Bottle—SC Johnson & Son, Inc.
Label—Peterson & Blyth Associates Inc.
Creative Director:
Chief Designer: Bottle—Diane Haworth/
Label—Jack Blyth
Client Firm: S. C. Johnson & Son Inc.

Entry: Command Performance
Category: Health and Beauty Aides
Award: Finalist 1985
Division: USA
Design Firm: Gregory Fossella Associates
Creative Director:
Chief Designer:
Client Firm:

Health and Beauty Aids 125

Entry	Vitalis
Category:	Health and Beauty Aids
Award:	Finalist 1987
Division:	USA
Design Firm:	Gerstman+Meyers
Creative Director:	Juan Concepcion
Chief Designer:	Karen Corell
Design Supervision	Richard Gerstman
Client Firm:	Bristol-Meyers

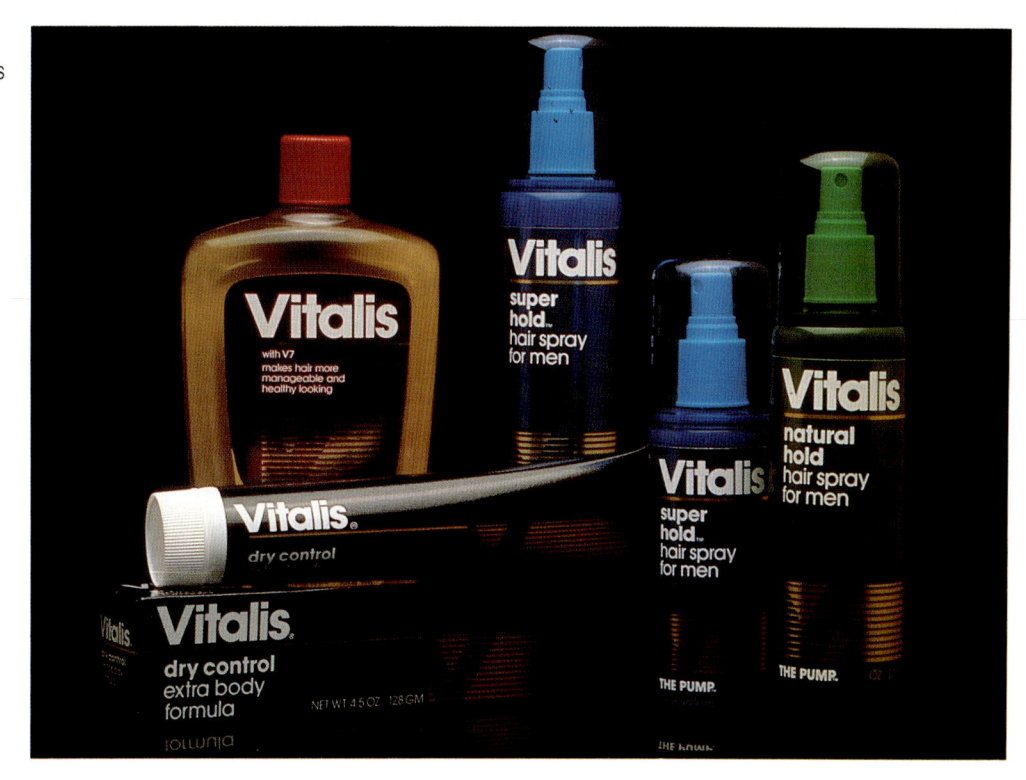

Entry	Jojoba Farms
Category:	Toiletries
Award:	Winner 1982
Design Firm:	Primo Angeli Graphics
Creative Director:	Primo Angeli
Chief Designer:	Primo Angeli, Ray Honda, Mark Jones, Shelly Weir
Client Firm:	Carme Inc.

126 Health and Beauty Aids

Entry	Prell Shampoo & Conditioner
Category:	Health and Beauty Aids
Award:	Finalist 1987
Division:	USA
Design Firm:	Libby Perszyk Kathman Inc.
Chief Designer:	Liz Kathman Grubow
Client Firm:	Procter & Gamble

Entry	Palmolive Shampoos
Category:	Health and Beauty Aids
Award:	Finalist 1987
Division:	International
Design Firm:	Colgate Corporate Design Group
Creative Director:	Dennis Brubaker
Chief Designer:	Sokie Lee
Client Firm:	Colgate-Palmolive/United Kingdom

Health and Beauty Aids 127

Entry	Toni Silkwave
Category:	Health and Beauty Aids
Award:	Finalist 1987
Division:	USA
Design Firm:	Coleman, Lipuma, Segal, & Morrill Inc.
Creative Director:	Edward Morrill
Chief Designer:	Ward Hooper
Client Firm:	The Gillette Co.

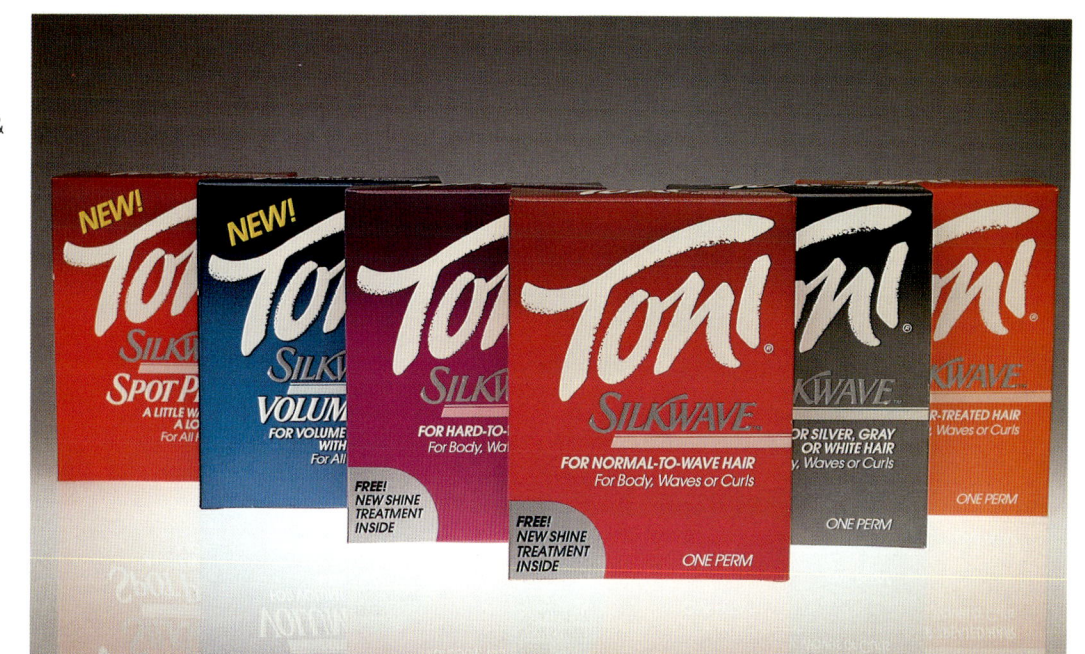

Entry	Toni Personal Perms
Category:	Health and Beauty Aids
Award:	Finalist 1987
Division:	USA
Design Firm:	Nason Design Associates, Inc.
Creative Director:	David Sandstrom, Charles Daigle
Chief Designer:	David Schwendeman
Client Firm:	Gillette Canada, Inc.

128 Health and Beauty Aids

Entry	Big Time
Category:	Beauty
Award:	Finalist 1987
Division:	International
Design Firm:	Hakuhodo Inc.
Creative Director:	Tokihiko Kimata, Hiroki Shiozu
Chief Designer:	Yasushi Miyamoto
Client Firm:	Kanebo Home Products Sales, Ltd.

Entry	Mink Difference Hair Products
Category:	Health and Beauty Aids
Award:	Finalist 1987
Division:	USA
Design Firm:	Nason Design Associates, Inc.
Creative Director:	David Sandstrom, Karin Kaplan
Chief Designer:	David Sandstrom, Steve DiFranza
Client Firm:	The Gillette Company/ Personal Care Division

Health and Beauty Aids 129

Entry	The Dry Look
Category:	Health and Beauty Aids
Award:	Finalist 1987
Division:	USA
Design Firm:	Coleman, Lipuma, Segal, & Morrill Inc.
Creative Director:	Edward Morrill
Chief Designer:	Edward Morrill
Client Firm:	The Gillette Co.

Entry	L'Envie Parfum Shampoos and Conditioners
Category:	Health and Beauty Aids
Award:	Finalist 1987
Division:	USA
Design Firm:	Peterson & Blyth Associates
Creative Director:	Ronald Peterson
Chief Designer:	Ronald Peterson
Client Firm:	S.C. Johnson & Son

130   Health and Beauty Aids

Entry	Pantene For Men
Category:	Health and Beauty Aids
Award:	Finalist 1987
Division:	USA
Design Firm:	Peterson & Blyth Associates
Creative Director:	Ronald Peterson
Chief Designer:	Ronald Peterson
Client Firm:	Richardson-Vicks

Entry	John Frieda Shampoos
Category:	Health and Beauty Aids
Award:	Finalist 1987
Division:	International
Design Firm:	Trickett & Webb
Chief Designer:	Lynn Trickett, Brian Webb & Fiona Skelsey
Client Firm:	John Frieda

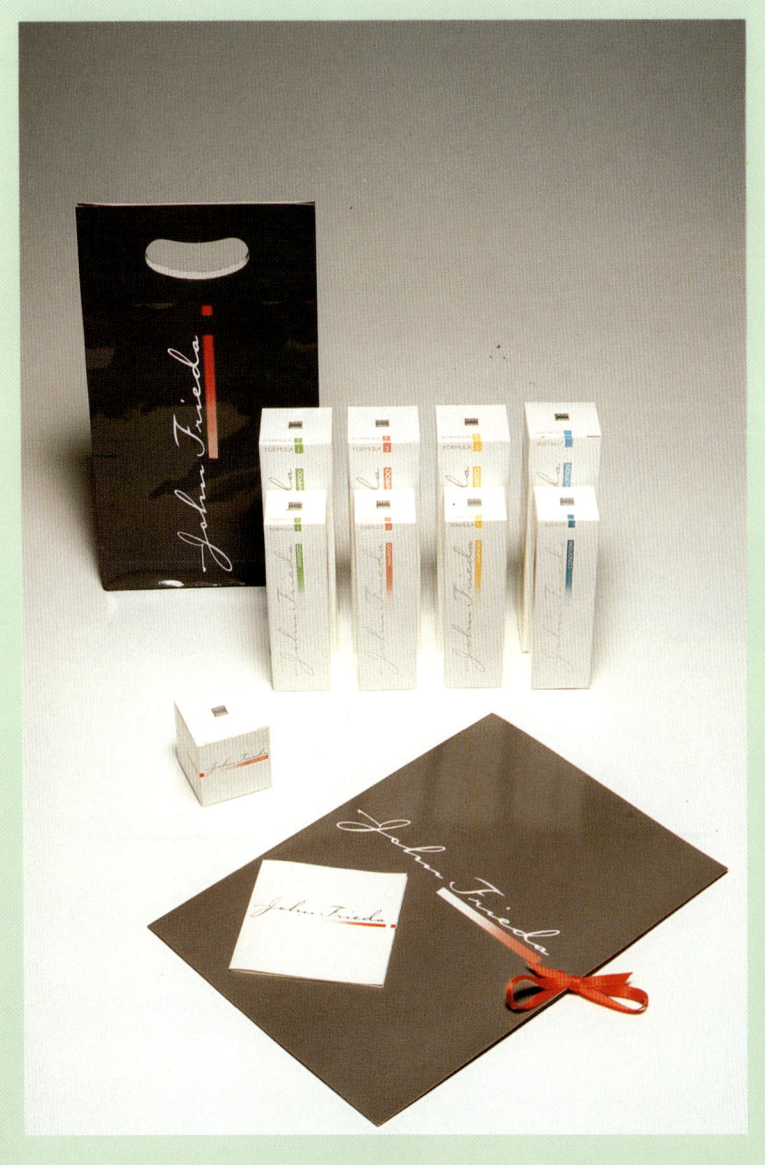

Health and Beauty Aids 131

Entry	Herbal Shampoo & Conditioner
Category:	Health and Beauty Aids
Award:	Finalist 1987
Division:	International
Design Firm:	Work Fabric Inc.
Creative Director:	Kaneto Yano
Chief Designer:	Yasuhiro Wakutsu
Client Firm:	Duskin Company Ltd.

Entry	Pantene Hair Care Products
Category:	Health and Beauty Aids
Award:	Finalist 1987
Division:	USA
Design Firm:	Peterson & Blyth Associates
Creative Director:	Ronald Peterson
Chief Designer:	Ronald Peterson
Client Firm:	Richardson-Vicks

132 Health and Beauty Aids

Entry	Jovan Liquid Creme Soaps
Category:	Toiletries
Award:	Winner 1981
Design Firm:	Robertz Webb & Co.
Creative Director:	Henry John Robertz
Chief Designer:	William Cagney
Client Firm:	Jovan, Inc.

Entry:	Shaklee Naturals
Category:	Health and Beauty Aids
Award:	Winner 1985
Division:	USA
Design Firm:	Primo Angeli, Inc.
Creative Director:	Primo Angeli
Chief Designer:	Ray Honda, Mark Jones
Client Firm:	Shaklee Corporation

Health and Beauty Aids 133

Entry	Vaseline Intensive Care Moisturizing Foam Baths
Category:	Health and Beauty Aids
Award:	Finalist 1987
Division:	USA
Design Firm:	Kollberg/Johnson Associates, Inc.
Client Firm:	Chesebrough-Pond's

Entry	Nivea Visage
Category:	Health and Beauty Aids
Award:	Finalist 1987
Division:	USA
Design Firm:	Si Friedman Associates, Inc.
Creative Director:	Si Friedman
Chief Designer:	Ken Goldner
Client Firm:	Biersdorf Inc.

Health and Beauty Aids

Entry	Maniac Pocket Perfumes
Category:	Cosmetics
Award:	Finalist 1987
Division:	USA
Design Firm:	Colgate Corporate Design Group
Creative Director:	Dennis Brubaker
Chief Designer:	Carolyn Pearle
Client Firm:	Colgate-Palmolive Enterprises

Entry	Etage Everyday Line
Category:	Cosmetics
Award:	Finalist 1987
Division:	USA
Design Firm:	Cato Desgrippes Beauchant Gobe
Creative Director:	Marc Gobe
Chief Designer:	Marc Gobe, Peter Levine
Client Firm:	The Brown Group—Etage

Health and Beauty Aids 135

Entry	Etage Evening Line
Category:	Cosmetics
Award:	Finalist 1987
Division:	USA
Design Firm:	Cato Desgrippes Beauchant Gobe
Creative Director:	Marc Gobe
Chief Designer:	Marc Gobe, Peter Levine
Client Firm:	The Brown Group—Etage

136 Health and Beauty Aids

Entry: Avon Summer Fruit Collection
Category: Specialty
Award: Finalist 1985
Division: USA
Design Firm: Avon
Creative Director:
Chief Designer:
Client Firm:

Entry Victoria's Secret
Category: Health and Beauty Aids
Award: Finalist 1987
Division: USA
Design Firm: Cato Desgrippes Beauchant Gobe
Creative Director: Marc Gobe
Chief Designer: Marc Gobe, Linda Sheintop
Client Firm: Victoria's Secret

Health and Beauty Aids 137

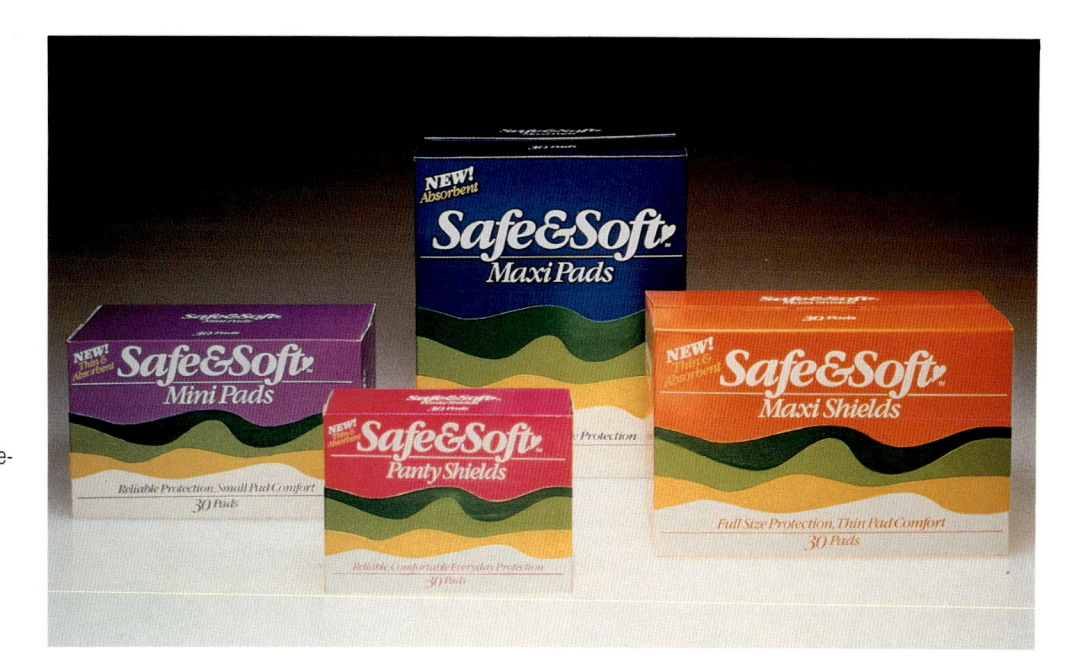

Entry	Safe 'N Soft
Category:	Pharmaceutical Over-The-Counter
Award:	Finalist 1985
Division:	USA
Design Firm:	The Berni Company
Creative Director:	Stephen H. Berni, PDC
Chief Designer:	Stuart M. Berni, FPDC
Client Firm:	Whitestone Products

Entry:	Avon Water Lily Collection
Category:	Health and Beauty Aids
Award:	Finalist 1985
Division:	USA
Design Firm:	Avon
Creative Director:	
Chief Designer:	
Client Firm:	

138 Health and Beauty Aids

Entry:	La Solaire
Category:	Over-The-Counter
Award:	Winner 1985
Division:	USA
Design Firm:	Schering Plough

Health and Beauty Aids 139

Entry	Luvs
Category:	Pharmaceutical Over-The-Counter
Award:	Finalist 1987
Division:	International
Design Firm:	Design Board/Behaeghel & Partners S.A.
Creative Director:	Thomas R. Salt
Chief Designer:	Denis Keller
Client Firm:	Procter & Gamble GmbH

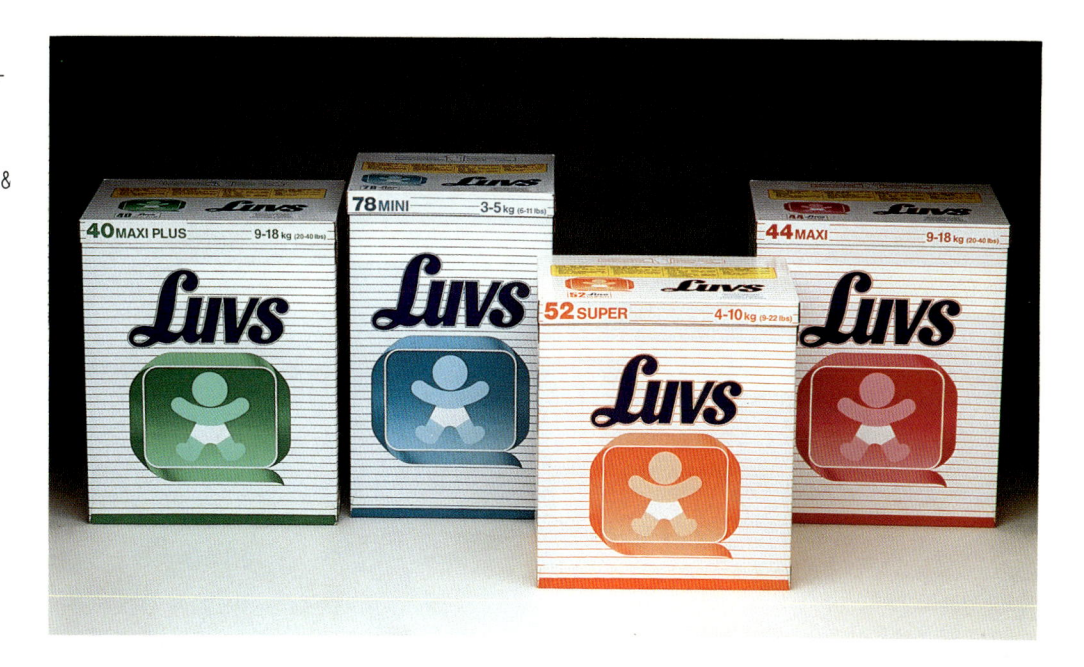

Entry:	Asda Nappies
Category:	Over-The-Counter
Award:	Finalist 1987
Division:	International
Design Firm:	Lock/Pettersen
Creative Director:	
Chief Designer:	
Client Firm:	

Health and Beauty Aids

Entry	AGA Enrich Skin Care Series
Category:	Health and Beauty Aids
Award:	Finalist 1987
Division:	International
Design Firm:	Work Fabric Inc.
Creative Director:	Kaneto Yano
Chief Designer:	Yasuhiro Wakutsu
Client Firm:	Duskin Company Ltd.

Entry	Cosmence
Category:	Cosmetics
Award:	Finalist 1987
Division:	International
Design Firm:	Beautiful Design House
Creative Director:	Olivier Desdoigts
Chief Designer:	Benedicte Gauquelin
Client Firm:	Club des Createurs de Beaute

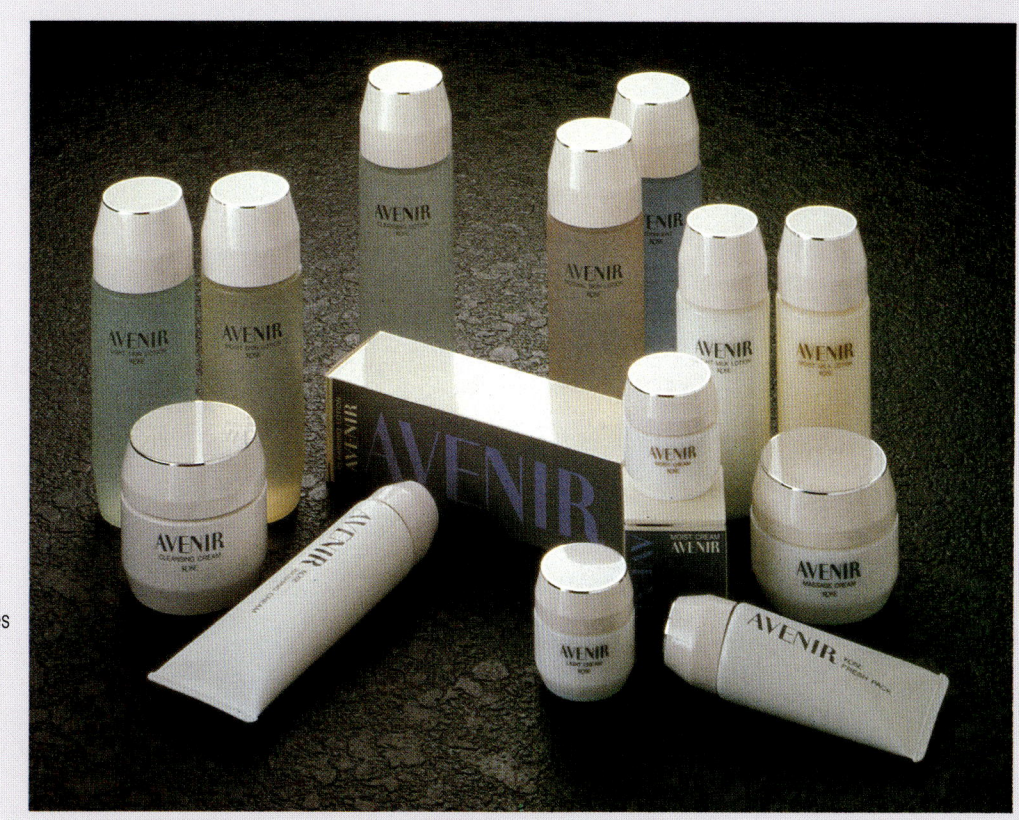

Entry	Kose Avenir Treatment Series
Category:	Health and Beauty Aids
Award:	Winner 1987
Division:	International
Design Firm:	Kobayashi Kose Co., Ltd.
Creative Director:	Haruyo Tahira
Chief Designer:	Chihiro Hayashi
Client Firm:	Kobayashi Kose Co., Ltd.

142 Health and Beauty Aids

Health and Beauty Aids 143

Entry:	Shiseido Cle De Peau
Category:	Cosmetics
Award:	Winner 1982
Design Firm:	Shiseido Co. Ltd./NY
Creative Director:	Yasuyuki Sawachi, Shunsaku Sugiura
Chief Designer:	Takashi Kaneko, Minoru Shiokawa
Client Firm:	Shiseido Co., Ltd.

Entry:	Albion Cell-Beauge
Category:	Cosmetics
Award:	Finalist 1987
Division:	International
Design Firm:	Albion Cosmetics Co./Design Room

Entry Beauty Mission
Category: Cosmetics
Award: Finalist 1987
Division: International
Design Firm: Avon Products Co., Ltd.
Creative Director: Timothy J. Musios
Chief Designer: Yoshitomo Ohama, Kazunori Umezawa
Client Firm: Avon Products Co., Ltd.

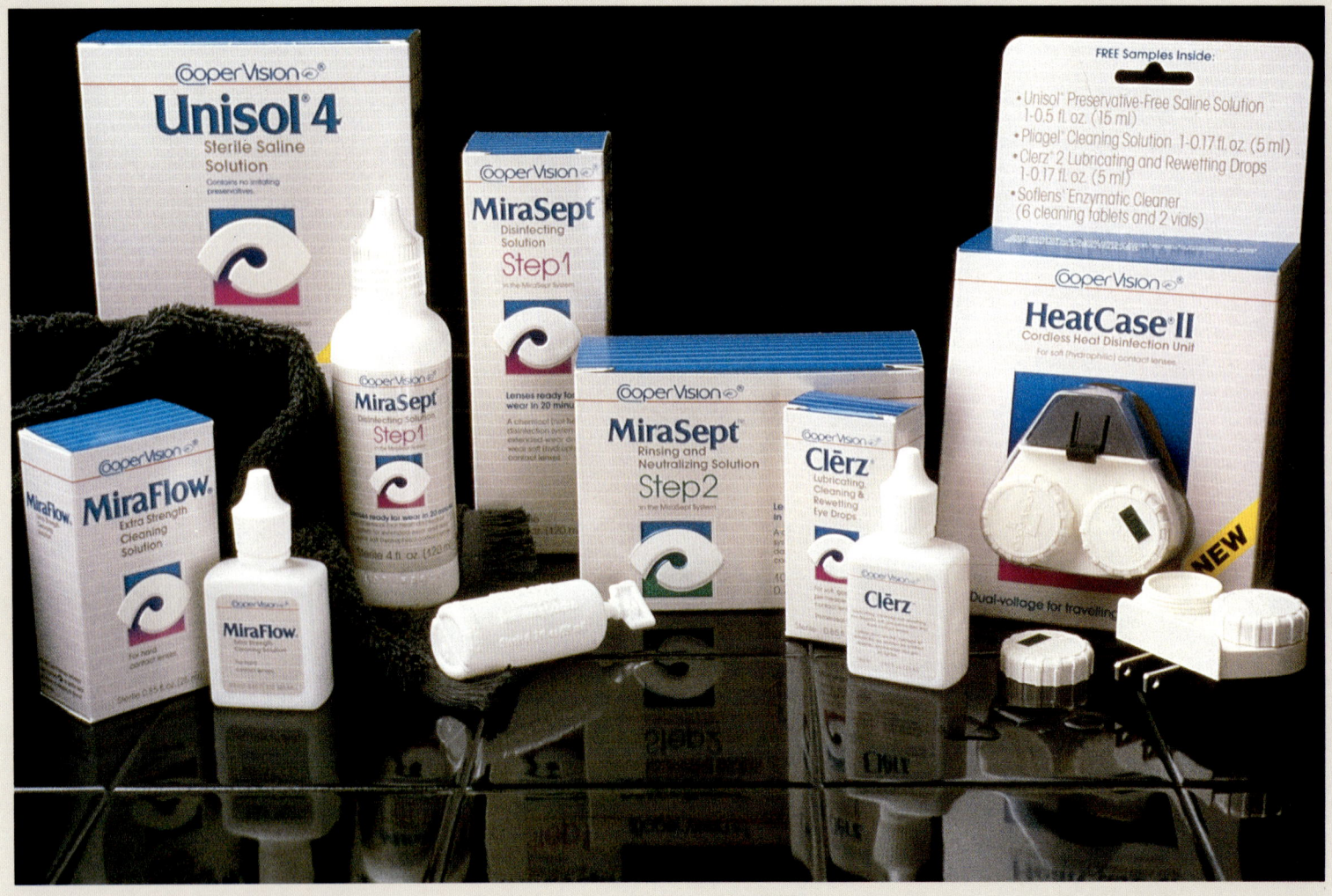

146 Health and Beauty Aids

Entry	Coopervision Eye Care System
Category:	Pharmaceutical Over-The-Counter
Award:	Finalist 1987
Division:	USA
Design Firm:	Landor Associates
Creative Director:	Kay Stout
Client Firm:	Coopervision Opthalmic Products

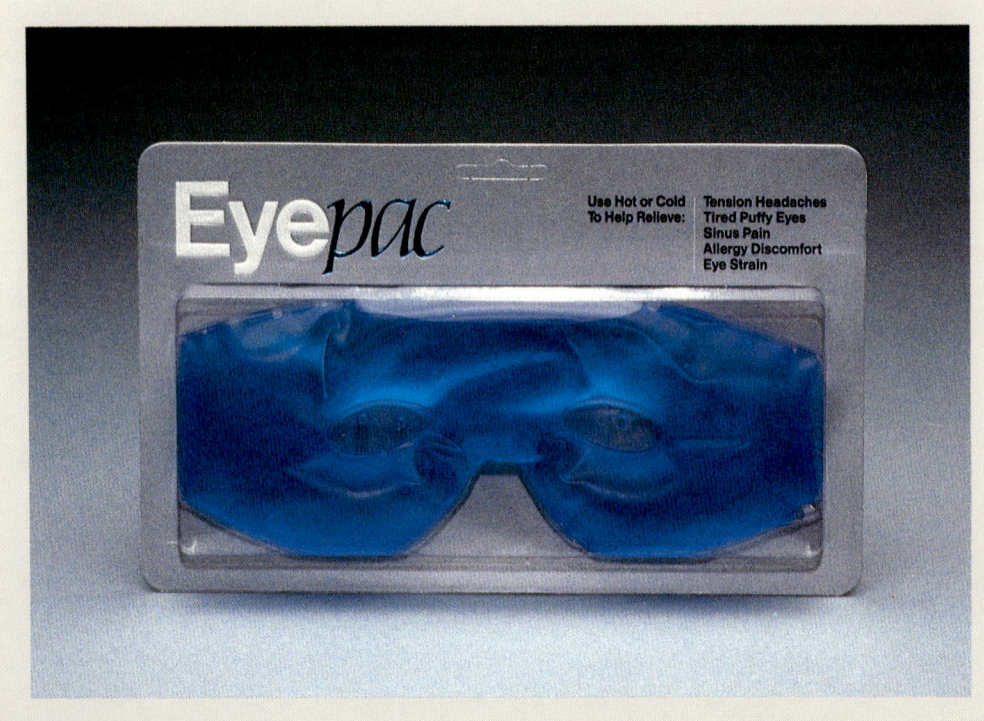

Entry	Eyepac
Category:	Pharmaceutical Over-The-Counter
Award:	Finalist 1987
Division:	USA
Design Firm:	Bagby Design Incorporated
Creative Director:	Steven Bagby
Chief Designer:	Jane Gittings
Client Firm:	Brighton Products

Health and Beauty Aids 147

Entry	Novahistine
Category:	Pharmaceutical Over-The-Counter
Award:	Finalist 1985
Division:	USA
Design Firm:	Landor Associates
✼✼Project Director:	Michael Livolsi
Creative Director:	Dick Young, Michael Livolsi
Client Firm:	Merrell Dow Pharmaceuticals, Inc.

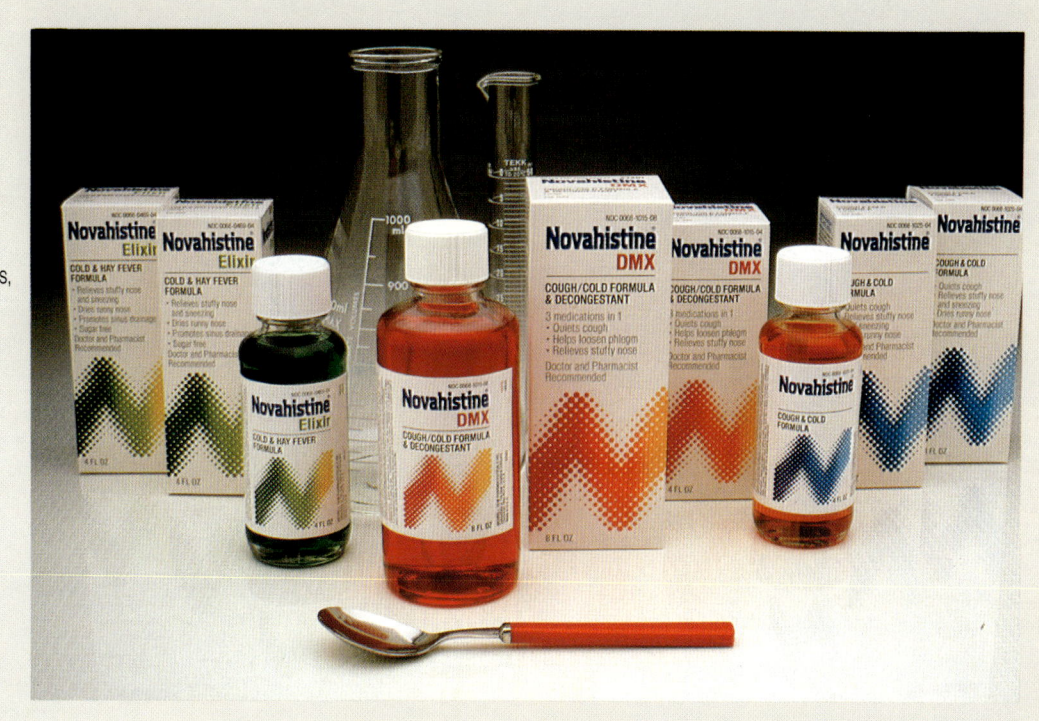

Entry	Thompson Premium Natural Pack
Category:	Pharmaceutical Over-The-Counter
Award:	Finalist 1985
Division:	USA
Design Firm:	Dennis S. Juett & Associates Inc.
Creative Director:	Dennis S. Juett
Chief Designer:	Dennis S. Juett
Client Firm:	Wm. T. Thompson Company

148   Health and Beauty Aids

Entry
Category: Pre-Pair, For-Play
Pharmaceutical Over-The-Counter
Award: Winner 1982
Design Firm: Bright & Associates
Creative Director: Keith Bright
Chief Designer: Raymond Wood
Client Firm: Trimensa Inc.

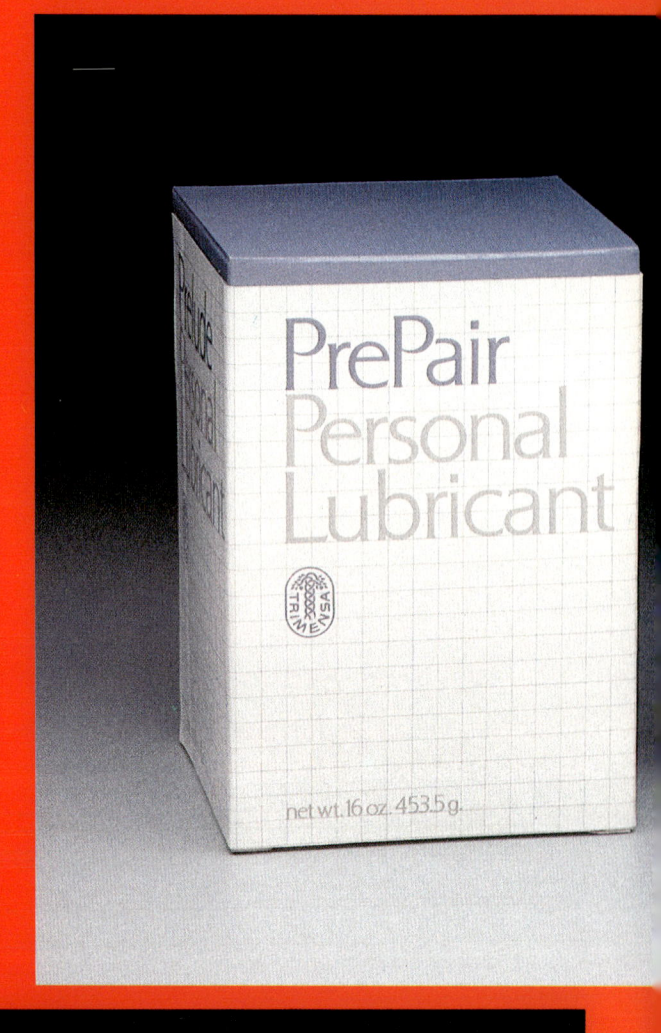

Health and Beauty Aids 149

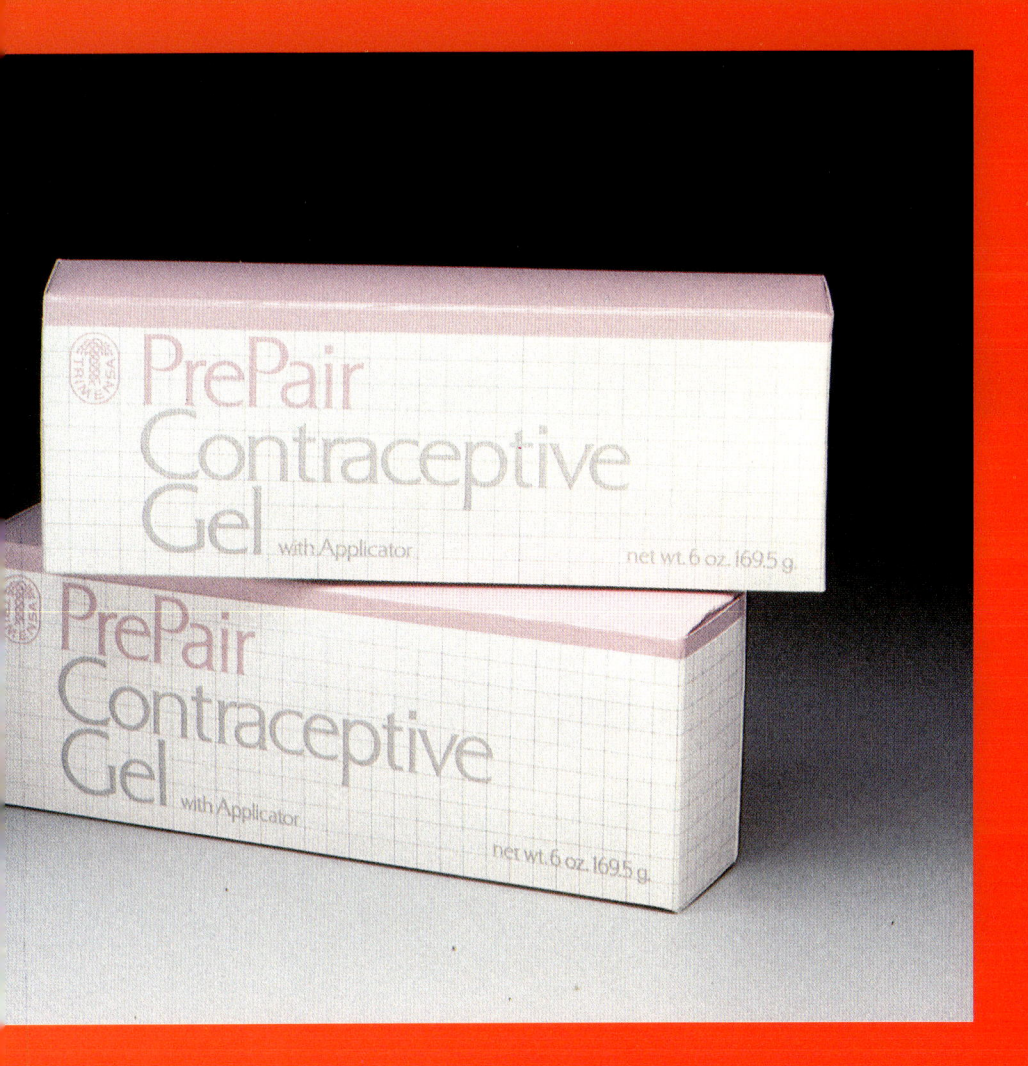

150　Health and Beauty Aids

Entry	Shiseido Revital
Category:	Cosmetics
Award:	Winner 1981
Design Firm:	Shiseido
Creative Director:	Shigeyoski Aoki
Chief Designer:	Shunsaku Sugiura & Tetsuo Togasawa
Client Firm:	Shisedo Co., Ltd.

Entry	MaxFactor 'fec'
Category:	Cosmetics
Award:	Winner 1987
Division:	International
Design Firm:	Taku Satoh Design Office
Creative Director:	Taku Satoh
Chief Designer:	Taku Satoh
Client Firm:	MaxFactor K.K.

Health and Beauty Aids 151

Entry	Boots Teenage Cosmetics
Category:	Cosmetics
Award:	Finalist 1987
Division:	International
Design Firm:	Lewis Moberly
Creative Director:	Mary Lewis
Chief Designer:	Jerome Berard
Client Firm:	The Boots Company PLC

Entry	Dimoda
Category:	Cosmetics
Award:	Finalist 1987
Division:	International
Design Firm:	Avon Products Co., Ltd.
Creative Director:	Timothy J. Musios
Chief Designer:	Yoshitomo Ohama, Kazunori Umezawa
Client Firm:	Avon Products Co., Ltd.

152 Health and Beauty Aids

Health and Beauty Aids 153

Entry Kose "BE" Makeup Series
Category: Cosmetics
Award: Winner 1985
Division: International
Design Firm: Kobayaski Kose Co.
Creative Director: Haruyo Tahira
Chief Designer: Hiroko Yamada
Client Firm: Kobayaski Kose Co.

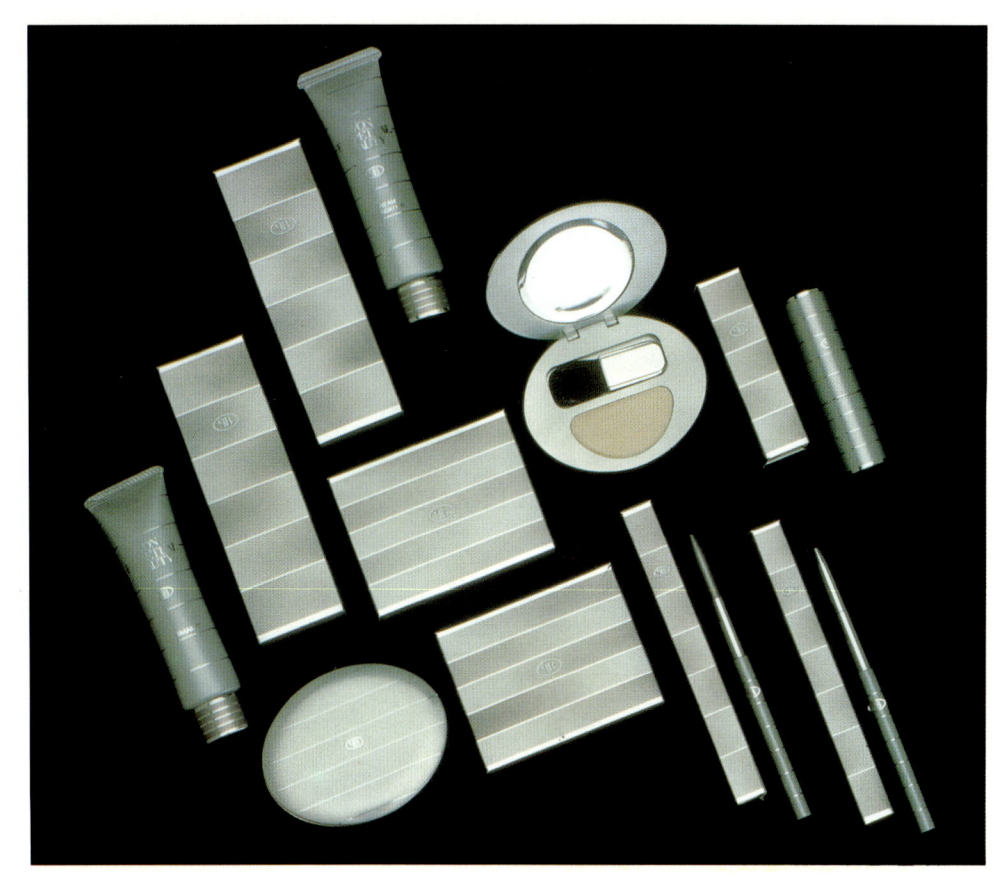

Entry: Avon Perpetual Beauty Line
Category: Cosmetics
Award: Finalist 1985
Division: International
Design Firm: Avon
Creative Director:
Chief Designer:
Client Firm:

Entry: Elegance
Category: Cosmetics
Award: Finalist 1987
Division: International
Design Firm: Albion Cosmetics Co./Design Room

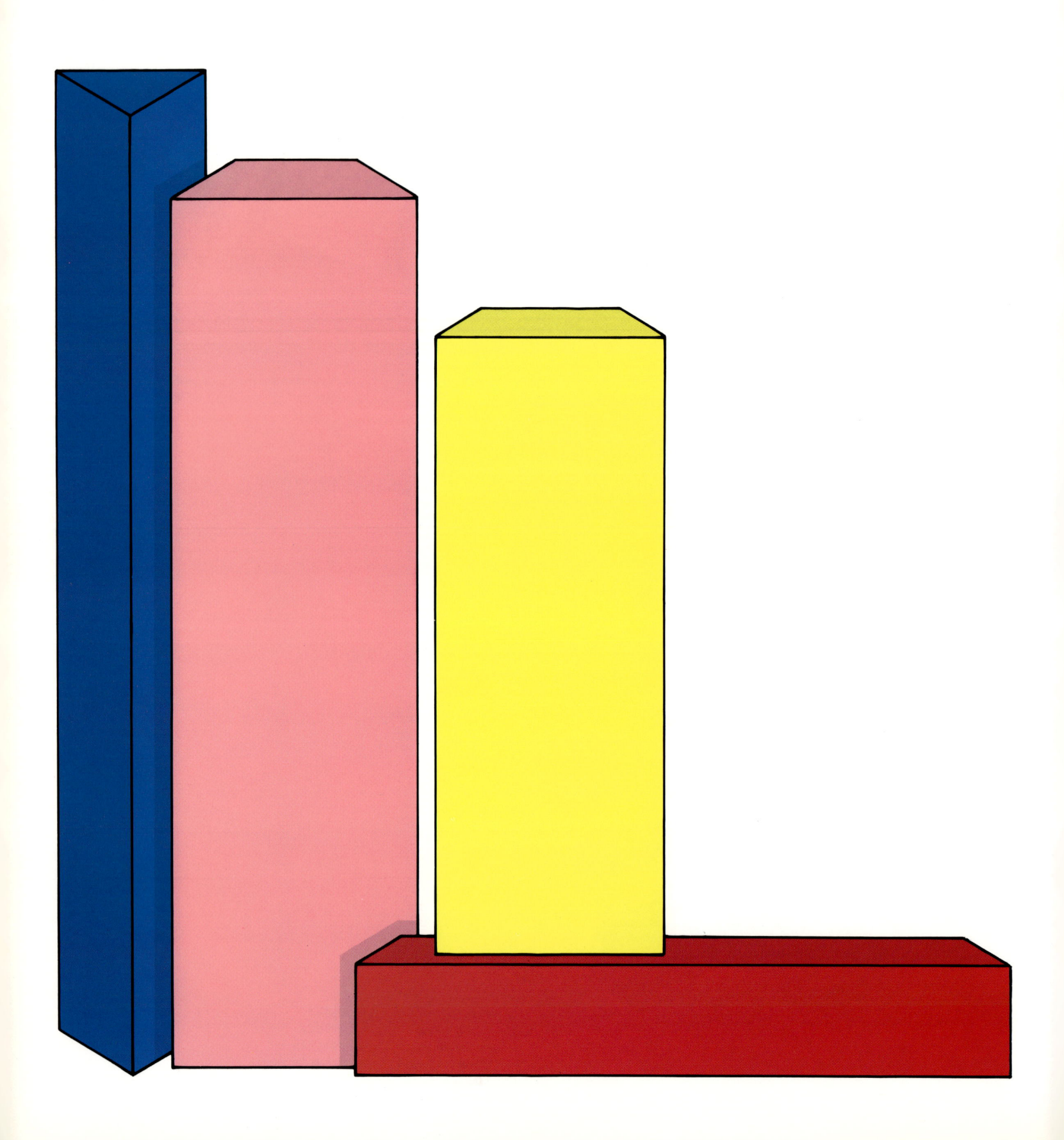

CHAPTER 4
Housewares

The winning entries in the PDC Gold Awards competition share a high level of creativity and a look that stands out among the competition in their respective categories. In each case, the designer exhibits a keen understanding of the market and the skill to create an image that is attractive to the consumer.

Breyer's, for example, destroyed the long-held belief that the color black and food packaging don't mix. By boldly and tastefully using black as a background, the designer presents the product—a mouth-watering dish of ice cream—with great impact.

The packaging for Clear Elegance offers a very strong and unique presentation of its product. By showing the clear cookware in use, the designer projects an image that appeals strongly to the consumer's taste.

The packaging for Okidata Microline marries the elegance of a black background with a highly technical line drawing of the product. Such a look offers a strong and clear communication of the quality and nature of the product.

Two other winners employ a more subtle approach. For Soumonka, the look is a refined one. The typography and graphics for the food packaging utilize a soft, delicate, multitextured look. The impact is a powerful one. Similarly, the solid package with a bold, elegant logo, subtle supporting copy and product description, makes the packaging for Shaklee Naturals a Gold Medal winner.

Barry Seelig
President/Creative Director
Apple Designsource Inc.

Housewares

Entry:	Visions
Category:	Housewares
Award:	Winner 1982
Design Firm:	Charles Biondo Design Associates
Creative Director:	Charles Biondo
Client Firm:	Corning Glass Works

Entry	Copperlon
Category:	Housewares
Award:	Finalist 1987
Division:	USA
Design Firm:	The Applebaum Company
Creative Director:	Harvy Applebaum
Chief Designer:	Harvy Applebaum, John Bechtold (Photographer)
Client Firm:	Benjamin & Medwin, Inc.

Housewares 157

Entry:	Pyrex
Category:	Housewares
Award:	Finalist 1987
Division:	USA
Design Firm:	Charles Biondo Design Associates
Creative Director:	Charles Biondo
Client Firm:	Corning Glass Works

Entry:	Clear Elegance
Category:	Housewares
Award:	Winner 1985
Division:	USA
Design Firm:	Charles Biondo Design Associates
Client Firm:	Corning Glass Works

158　Housewares

Entry	Sensation Dinnerware
Category:	Housewares
Award:	Finalist 1987
Division:	USA
Design Firm:	RH Macy Corporate Graphics
Creative Director:	Vincent Dimino
Client Firm:	RH Macy & Company, Inc.

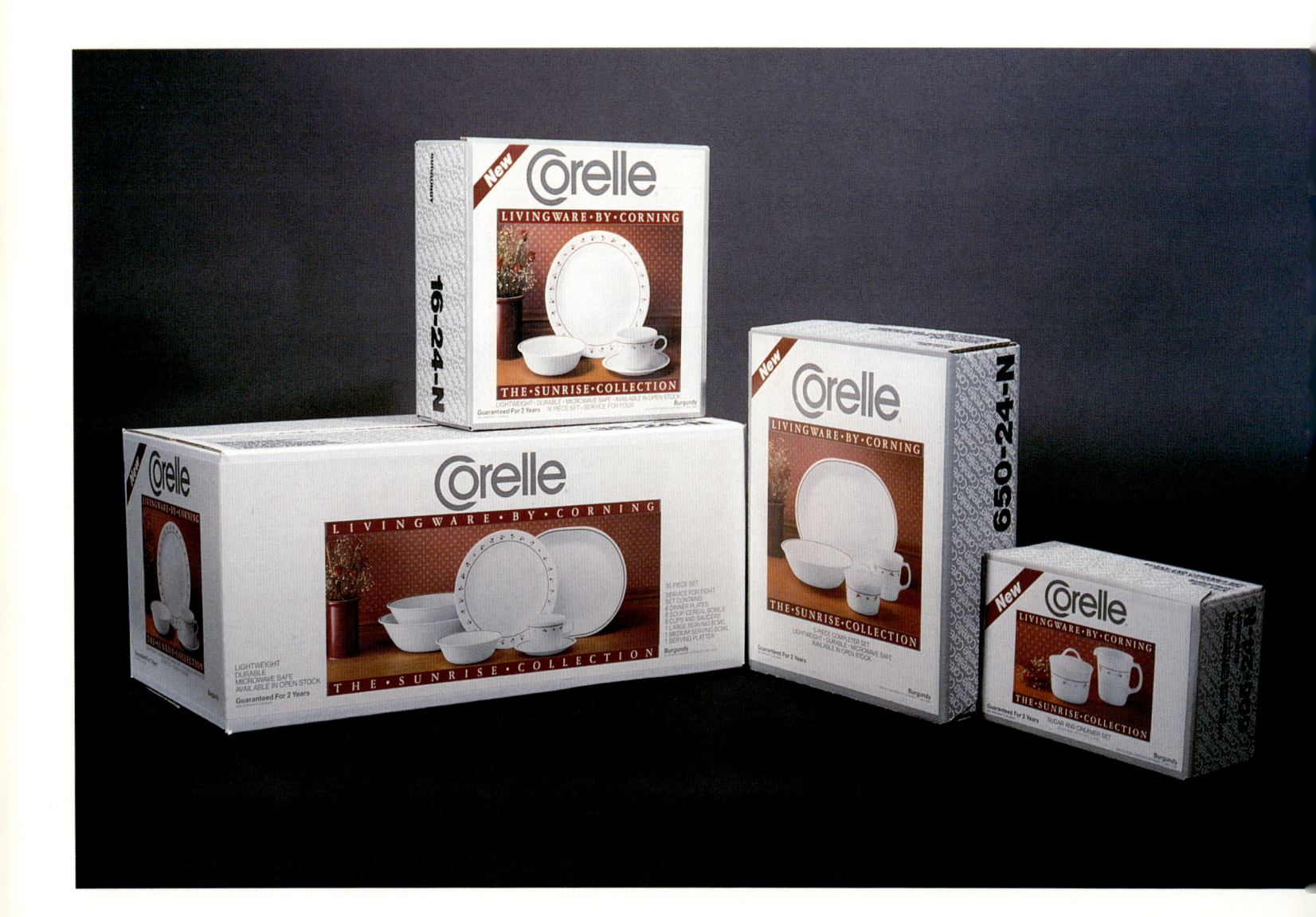

Entry	Spong Logic
Category:	Housewares
Award:	Finalist 1987
Division:	International
Design Firm:	Trickett & Webb
Chief Designer:	Lynn Trickett, Brian Webb & Penny Sykes
Client Firm:	Spong PLC

Entry:	Corelle
Category:	Housewares
Award:	Finalist 1985
Division:	USA
Design Firm:	Source/Inc.
Creative Director:	William J. O'Connor
Chief Designer:	Charles Cybul, Hedy Lim Wong
Client Firm:	Corning Glass Works

160 Housewares

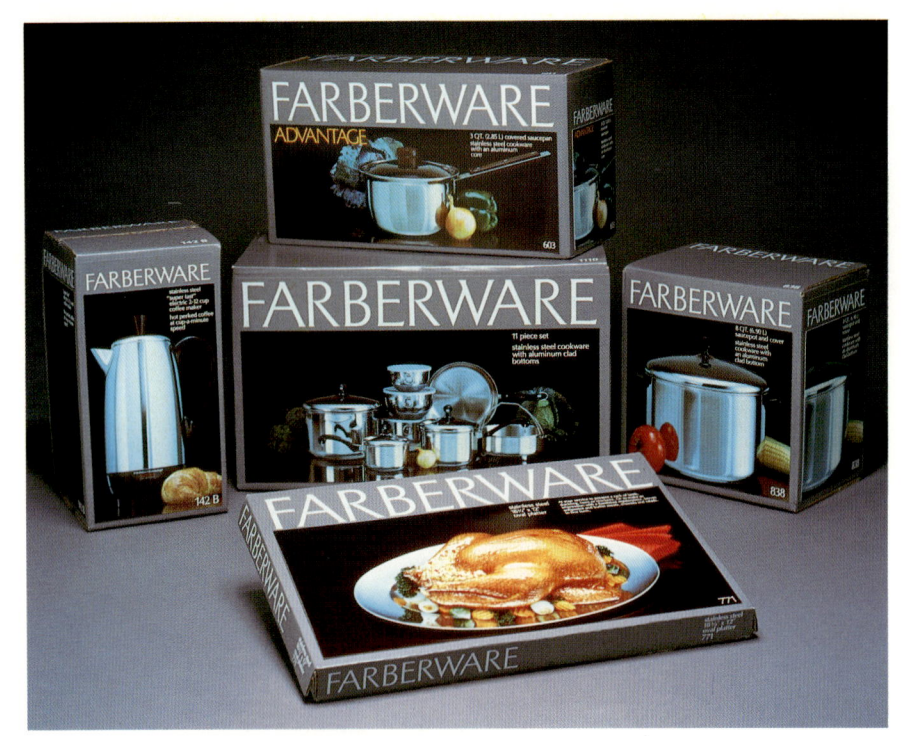

Entry	Farberware
Category:	Housewares
Award:	Winner 1985
Division:	USA
Design Firm:	Gerstman+Meyers Inc.
✣✣Design Supervision:	Herbert M. Meyers
Creative Director:	Juan Concepcion
Chief Designer:	Larry Riddell
Client Firm:	Farberware Inc.

Entry:	Cutlery Sets
Category:	Housewares
Award:	Winner 1981
Design Firm:	JC Penny In House—Creative Services

Entry:	Smythe & Cook
Category:	Housewares
Award:	Finalist 1985
Division:	USA
Design Firm:	The Pushpin Group Inc.
Creative Director:	Seymour Chwast
Chief Designer:	Seymour Chwast, Michael Aron
Client Firm:	Associated Merchandising Corporation

Entry:	Show It Alls
Category:	Household
Award:	Winner 1980
Design Firm:	Charles Biondo Design Associates
Creative Director:	Charles Biondo
Client Firm:	Corning Glass Works

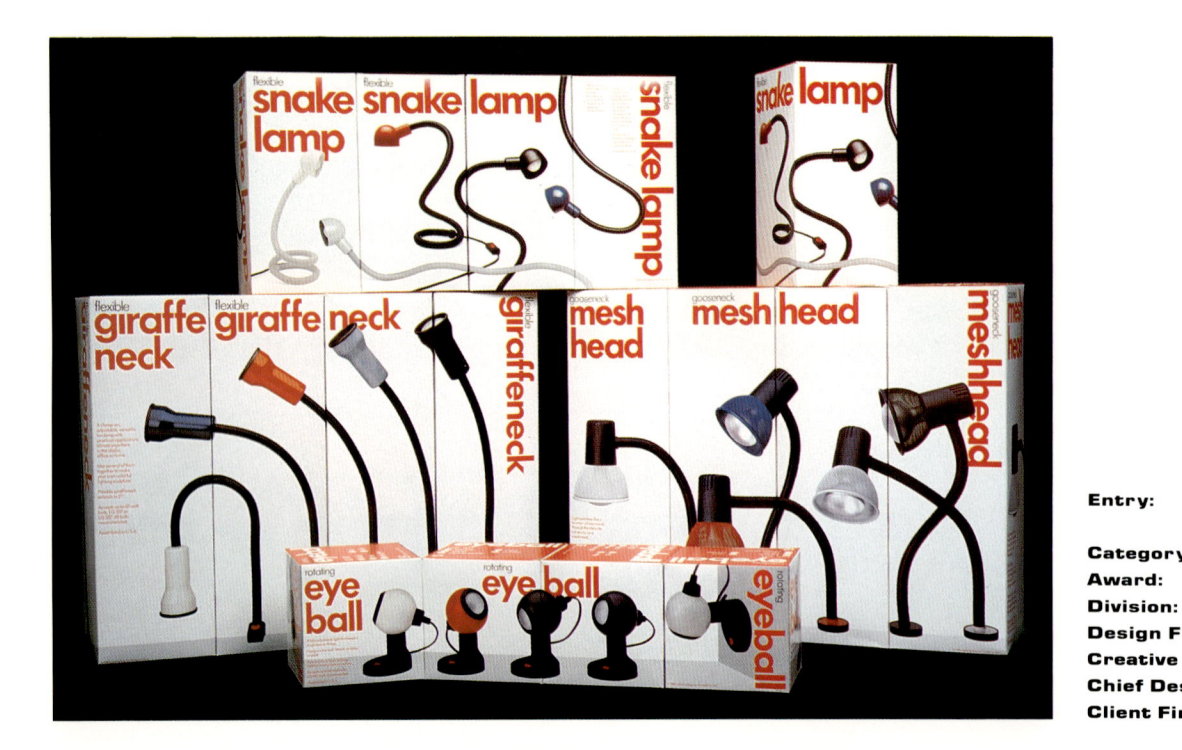

Housewares 162

Entry:	Giraffe Mesh/Snake/Eyeball Lighting
Category:	Housewares
Award:	Finalist 1985
Division:	USA
Design Firm:	Porter/Matjasich & Associates
Creative Director:	Allen Porter
Chief Designer:	Allen Porter, Lisa Ellert
Client Firm:	Lightning Bug, Ltd.

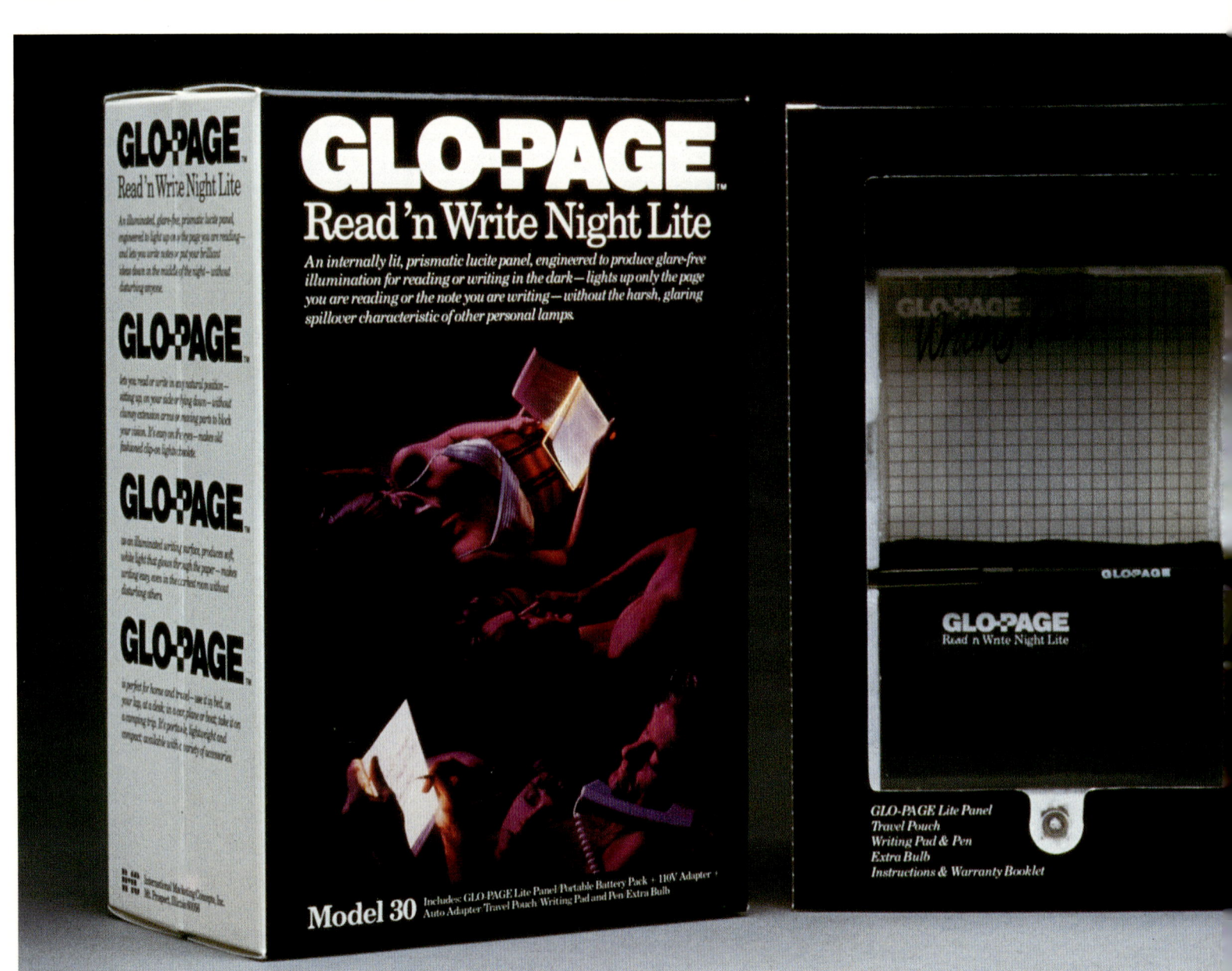

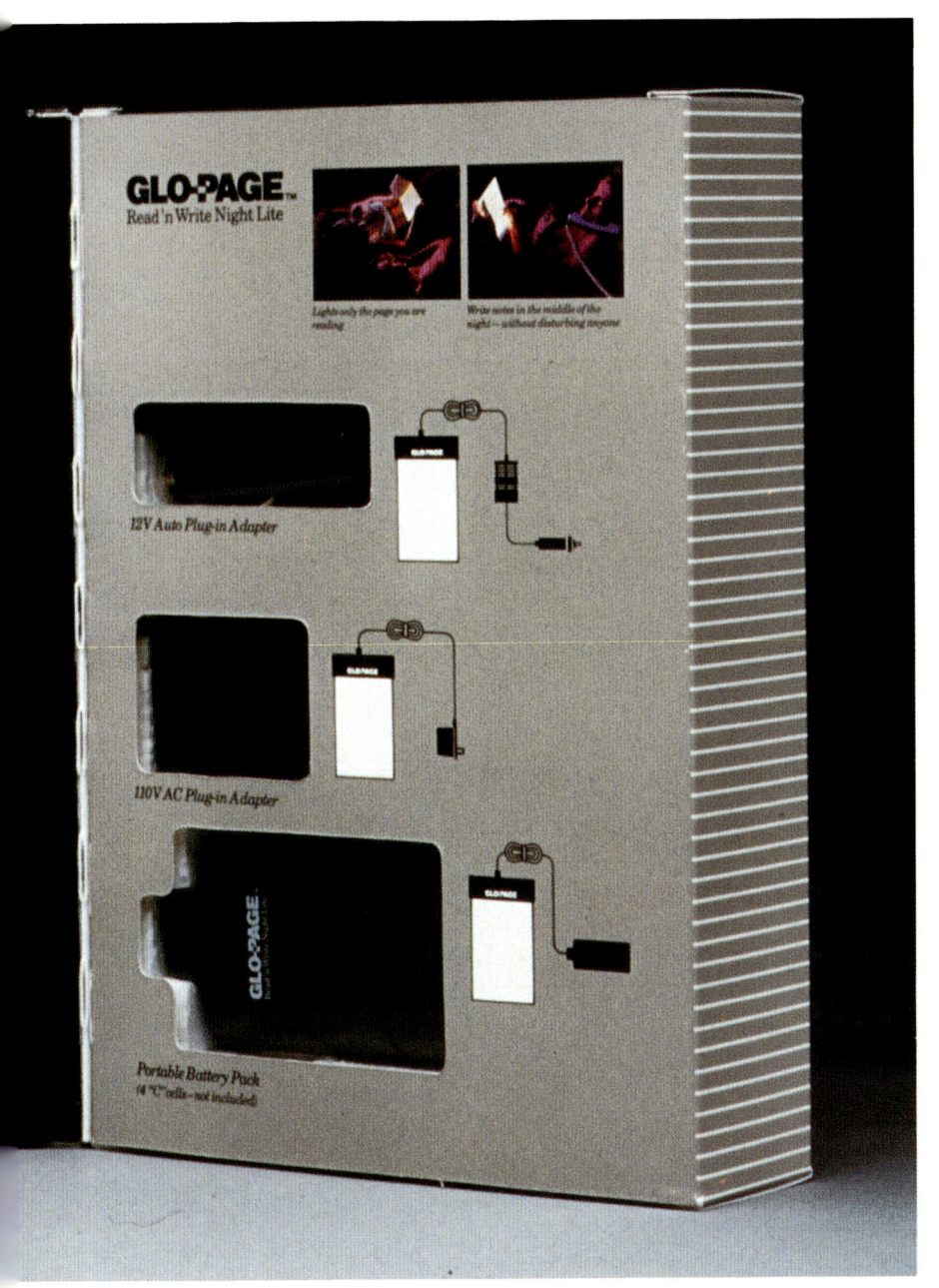

Housewares 163

Entry	Norelco Packaging System
Category:	Housewares
Award:	Finalist 1987
Division:	USA
Design Firm:	King Casey, Inc.
Creative Director:	Eugene J. Casey, Jr.
Chief Designer:	William Miller
Client Firm:	North American Philips Corporation

Entry	Glo-Page Reading Light
Category:	Housewares
Award:	Finalist 1987
Division:	USA
Design Firm:	Porter Matjasich & Associates
Creative Director:	Allen Porter
Chief Designer:	Allen Porter, Louis Claps
Client Firm:	International Marketing Concepts, Inc.

164 Housewares

Entry	Krone Telephone
Category:	Housewares
Award:	Finalist 1985
Division:	USA
Design Firm:	Gaylord Adams/Don Flock & Assoc.
Creative Director:	Gaylord Adams
Chief Designer:	Gaylord Adams
Client Firm:	Krone U.S.A. & GMBH

Entry:	Shower Massage
Category:	Housewares
Award:	Finalist 1985
Division:	USA
Design Firm:	Charles Biondo Design Associates
Client Firm:	Teledyne Waterpik

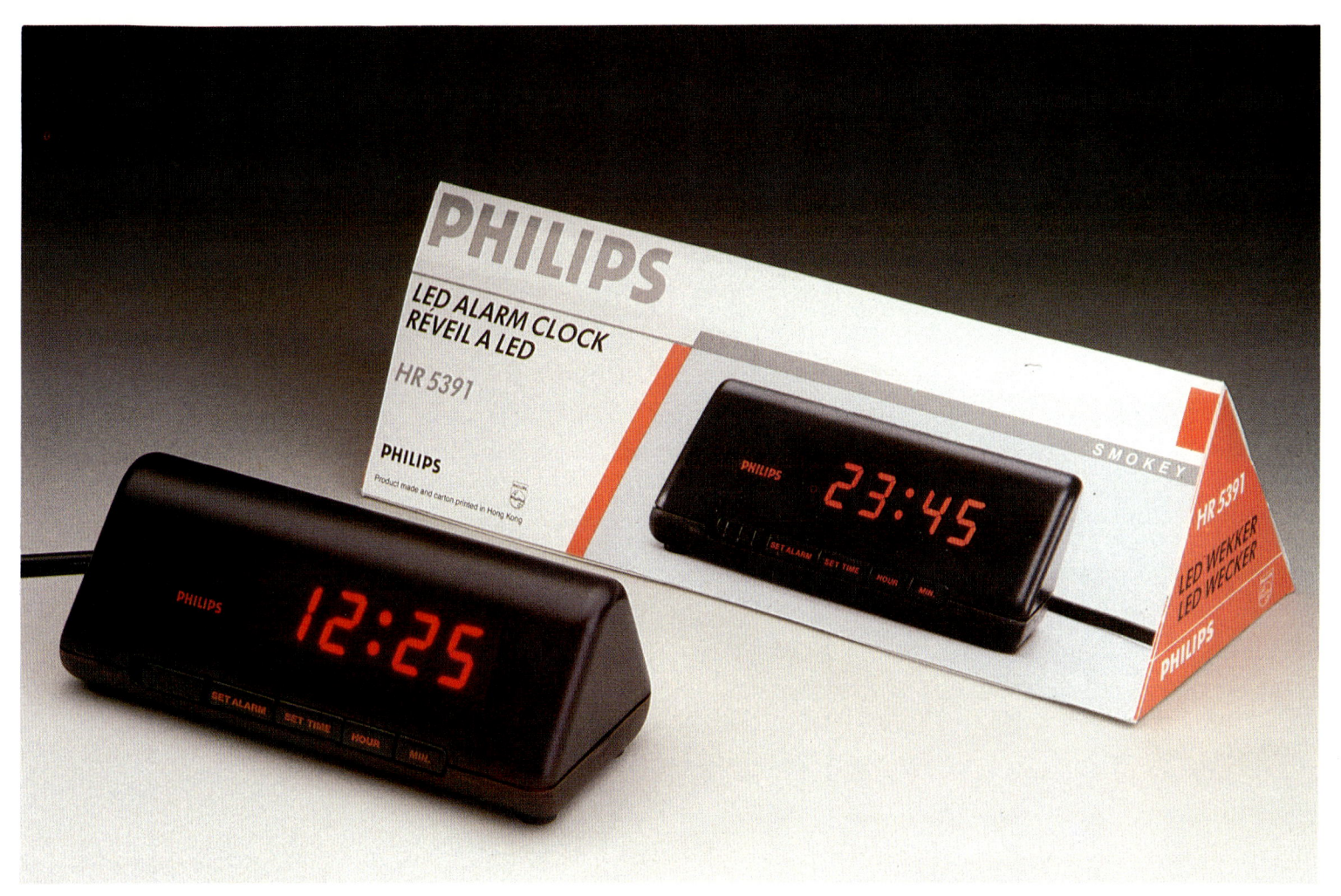

Housewares 165

Entry	Philips LED Alarm Clock HR 5391
Category:	Housewares
Award:	Finalist 1987
Division:	International
Design Firm:	Ad Broeders Graphic Design B.V.
Creative Director:	Ad Broeders
Chief Designer:	Ad Broeders
Client Firm:	Philips International B.V.

Entry	Dorma/Jenny Wren
Category:	Housewares
Award:	Finalist 1987
Division:	International
Design Firm:	Trickett & Webb
Chief Designer:	Lynn Trickett, Brian Webb & Colin Sands
Client Firm:	Dorma/Brinkhaus

CHAPTER 5
Sports and Leisure

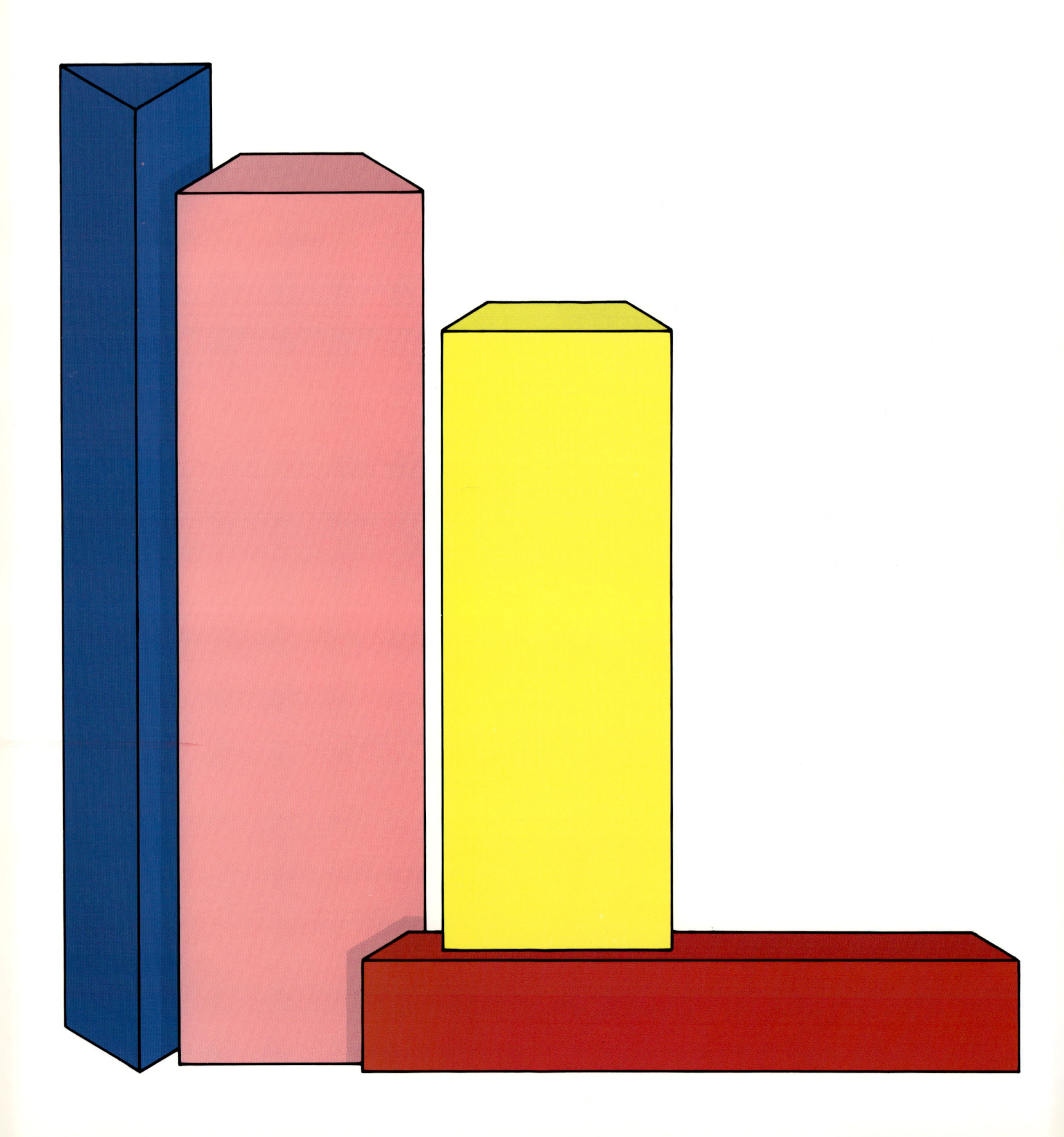

The PACKAGE DESIGN COUNCIL 1985 Gold Award Competition was a black year in the best sense of that word. It was the year that conventional wisdom about the folly of black packages was finally laid to rest, with a solid and resounding thud, across a convincingly broad and diverse range of product categories.

Parker Brothers commemorated fifty years of MONOPOLY with a striking black metal cannister; Corning showcased the clarity of CLEAR ELEGANCE cookware with an elegantly simple black carton; BREYERS proved that a black ice cream package not only could work, it could out-class the best of the superpremium competition; and LA SOLAIRE used high key, high gloss black packaging to bring a higher sense of style and a higher price point to the sun tanning products shelf.

In 1985, Japanese designers reaffirmed for their American and European colleagues the cultural roots of their minimalist visual design idiom. This was exemplified by the modulation of simple geometry, minimal color and volumes of empty space in Japanese designers' work for KOSE 'BE' SERIES cosmetics and SOUMONKA confections.

The 1985 competition was also a study in contrasts and offered conclusive evidence that brand personalities are clearly rooted in each product's design and trade dress. From the arts and crafts inspired graphics of Erlanger Beer, whose package design and REINHEITSGEBOT label story were part of a strategy to take the brand from a superpremium market position to that of a specialty beer, to the evocative power of the new WELCH'S ORCHARD graphics system which captured and synthesized so effectively all of the notions of wholesome country freshness that consumers have always had about this brand, package design made substantial differences.

The OKIDATA packaging system worked so well to clearly inform prospects and customers about the specific cartridge and ribbon inside, that prospects and customers were given positive impressions about the performance of the products themselves. From the functional linear business look of OKIDATA and FARBERWARE to the evocative typography of POSADA and SHAKLEE NATURAL, MOUSSE CHEZ NOUS and MICHEL GUERARD BELGIAN CHOCOLATES, PDC's 1985 Gold Award Competition demonstrated the finesse and visual variety with which accomplished designers can make substantial distinctions among a store full of me-too products.

William J. O'Connor
Executive Vice President
Source/Inc.
Juror—Package Design Council
International's 1985 Gold
Awards Competition

168 Sports and Leisure

Entry	Jenga/Game of Skill
Category:	Sports/Leisure
Award:	Finalist 1987
Division:	USA
Design Firm:	Nason Design Associates, Inc.
Creative Director:	David Sandstrom, Alec Jutsum
Chief Designer:	David Schwendeman
Client Firm:	Milton Bradley Company

Sports and Leisure 169

Entry	Rubik's Magic
Category:	Sports/Leisure
Award:	Finalist 1987
Division:	USA
Design Firm:	Source/Inc.
Creative Director:	Charles Cybul
Chief Designer:	Bernard Dolph, Hideo Watanabe
Client Firm:	Matchbox Toys (USA) Ltd.

Entry:	Monopoly 1935 Commemorative Edition
Category:	Recreation
Award:	Winner 1985
Division:	USA
Design Firm:	Parker Brothers

170 Sports and Leisure

Entry	Siku Trucks
Category:	Sports/Leisure
Award:	Finalist 1987
Division:	USA
Design Firm:	Irv Koons Associates, Inc.
Creative Director:	Irv Koons
Chief Designer:	Paul Gee
Client Firm:	Siku of America/Sieper Werke GmBh

Entry	Learning Curves
Category:	Sports/Leisure
Award:	Finalist 1987
Division:	USA
Design Firm:	Gerstman+Meyers
Creative Director:	Juan Concepcion
Chief Designer:	Karen Corell
Design Supervision:	Herbert M. Meyers
Client Firm:	Panosh Place

Sports and Leisure 171

AH-1S COBRA ATTACK HELICOPTER

Monogram 1/48 SCALE MODEL KIT • MODELE REDUIT

Entry	Monogram Models
Category:	Sports/Leisure
Award:	Finalist 1987
Division:	USA
Design Firm:	Joss Design Group
Chief Designer:	Gwen Moy
Client Firm:	Monogram Models, Inc.

Entry	Early American & Victorian Dolls
Category:	Specialty
Award:	Finalist 1987
Division:	USA
Design Firm:	Avon Products
Creative Director:	Tim Musios
Chief Designer:	Michael Mammina
Client Firm:	Avon Products

172 Sports and Leisure

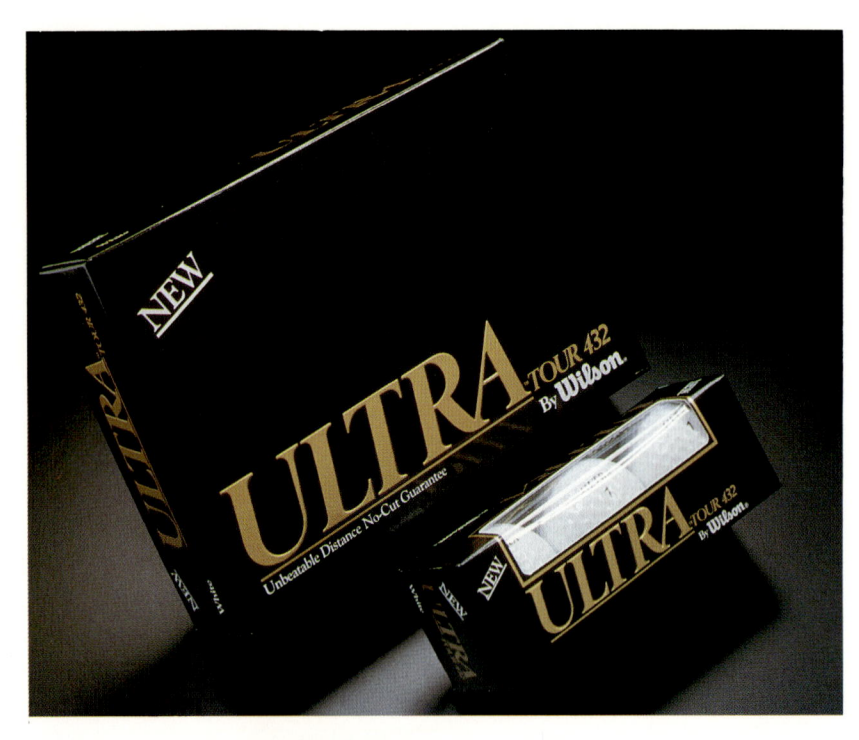

Entry:	Mind Shadow
Category:	Recreation
Award:	Finalist 1985
Division:	USA
Design Firm:	Sidjakov Berman Gomez & Partners
Creative Director:	Nicolas Sidjakov & Jerry Berman
Chief Designer:	James Nevins & Jackie Foshaug
Client Firm:	Activision

Entry:	Ultra Tour 432 Golf Balls
Category:	Sport
Award:	Finalist 1985
Division:	USA
Design Firm:	Wallner, Harbauer, Bruce & Associates
Creative Director:	George Bruce
Chief Designer:	Marie V. Grygienc
Client Firm:	Wilson Sporting Goods

Entry	Ektelon Eyewear
Category:	Sports/Leisure
Award:	Finalist 1987
Division:	USA
Design Firm:	Gamble and Bradshaw Design
Creative Director:	Chet Makoski
Chief Designer:	Linda Goldsmith
Client Firm:	Ektelon

MINDSHADOW™

AN ILLUSTRATED TEXT ADVENTURE

ActiVision

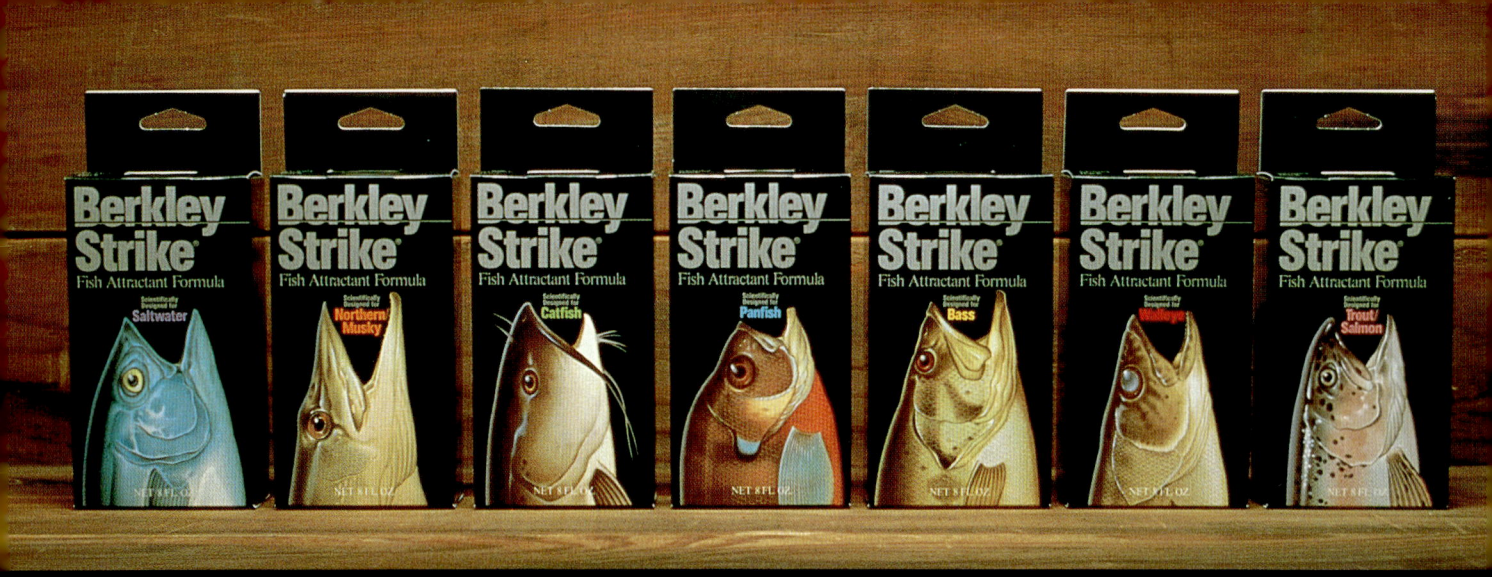

Entry	Berkley Outdoor Products
Category:	Sports/Leisure
Award:	Winner 1987
Division:	USA
Design Firm:	Sidjakov Berman Gomez & Partners
Creative Director:	Nicolas Sidjakov, Jerry Berman
Chief Designer:	Jackie Foshaug, Courtney Reeser
Client Firm:	Berkley

Entry	Berkley Strike
Category:	Sports/Leisure
Award:	Finalist 1987
Division:	USA
Design Firm:	Sidjakov Berman Gomez & Partners
Creative Director:	Nicolas Sidjakov, Jerry Berman
Chief Designer:	Barbara Vick, Courtney Reeser
Client Firm:	Berkley

Entry	Year 'Round Calendar
Category:	Specialty
Award:	Finalist 1987
Division:	USA
Design Firm:	Florville Design & Analysis, Inc.
Creative Director:	Patrick Florville
Chief Designer:	Patrick Florville
Client Firm:	Florville Design & Analysis, Inc.

Entry	Prime Fishing Line
Category:	Sports/Leisure
Award:	Finalist 1987
Division:	USA
Design Firm:	Peterson & Blyth Associates
Creative Director:	David Scarlett
Chief Designer:	Peterson & Blyth Associates
Client Firm:	Du Pont

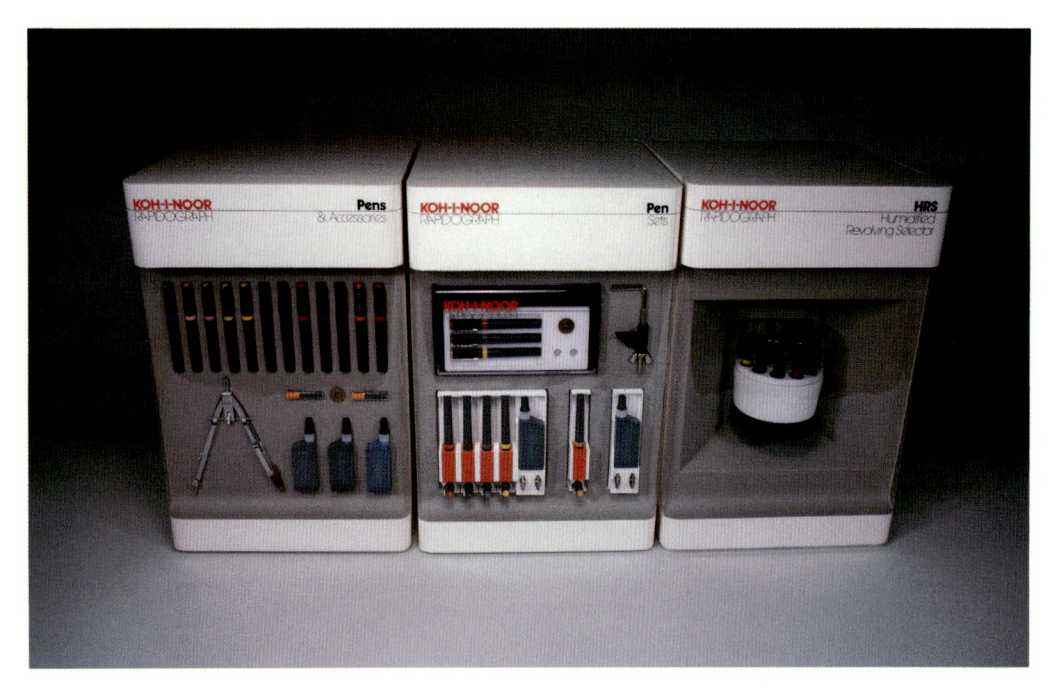

176 Sports and Leisure

Entry: Koh-I-Noor Rapidograph Pens
Category: Leisure Time Products
Award: Winner 1981
Design Firm: Robert Hain Associates

Entry: Bucilla Needlecraft Kits
Category: Leisure Time Products
Award: Winner 1982
Design Firm: Robert Hain Associates

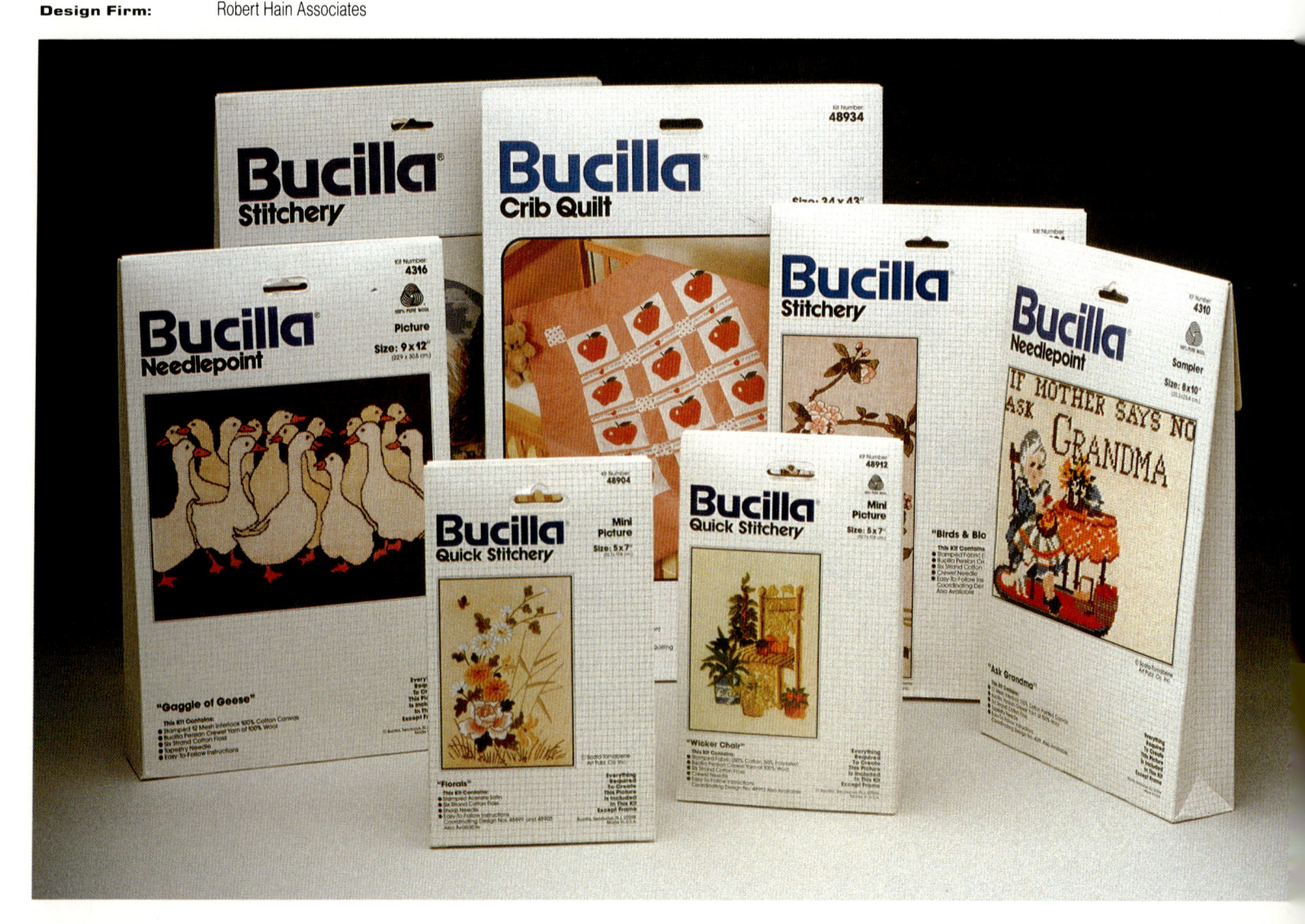

Sports and Leisure 177

Entry:	Prang Art Cases
Category:	Recreational
Award:	Finalist 1985
Division:	USA
Design Firm:	Robert Hain Assoc.

178 Sports and Leisure

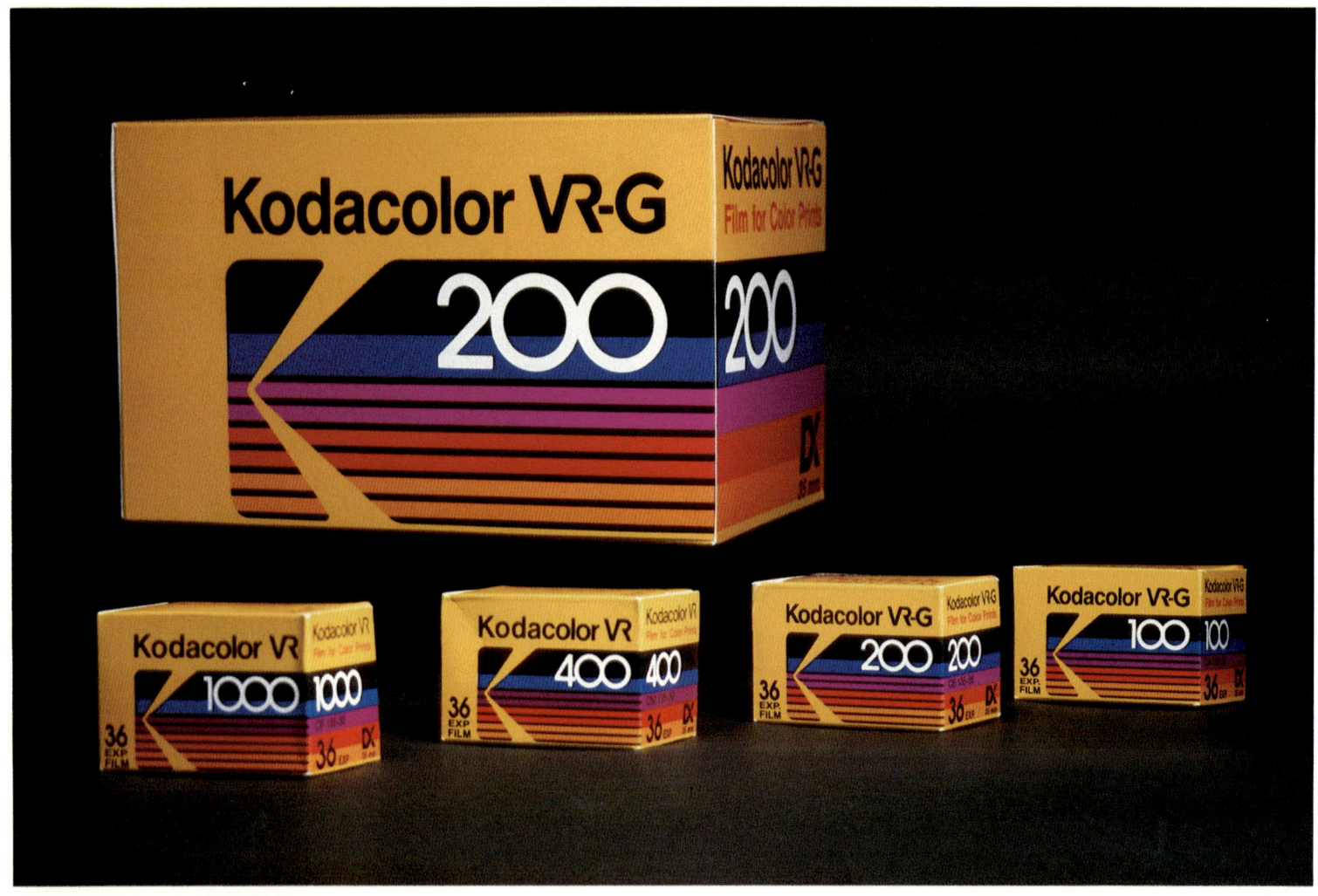

Entry	Kodak VRG
Category:	Sports/Leisure
Award:	Finalist 1987
Division:	USA
Design Firm:	Selame Design
Creative Director:	Dave Adams
Chief Designer:	Joseph Selame
Client Firm:	Eastman Kodak Company

Entry	Easy Shot 1, 2, 2AF Series
Category:	Sports/Leisure
Award:	Finalist 1987
Division:	USA
Design Firm:	Ronald Emmerling Design Inc.
Chief Designer:	Ronald Emmerling
Client Firm:	Keystone Camera Corporation

Entry	Agatha Christie Murderer's Row Collection
Category:	Sports/Leisure
Award:	Finalist 1987
Division:	USA
Design Firm:	Designed To Print, Inc.
Creative Director:	Tree Trapanese/Peggy Leonard
Chief Designer:	Tree Trapanese/Peggy Leonard
Client Firm:	HBO/Cannon Video

Entry	Le Clic 35MM Camera Line
Category:	Sports/Leisure
Award:	Finalist 1987
Division:	USA
Design Firm:	Ronald Emmerling Design Inc.
Creative Director:	Ronald Emmerling
Chief Designer:	Dale Clark
Client Firm:	Le Clic Products Inc.

CHAPTER 6
Hardware

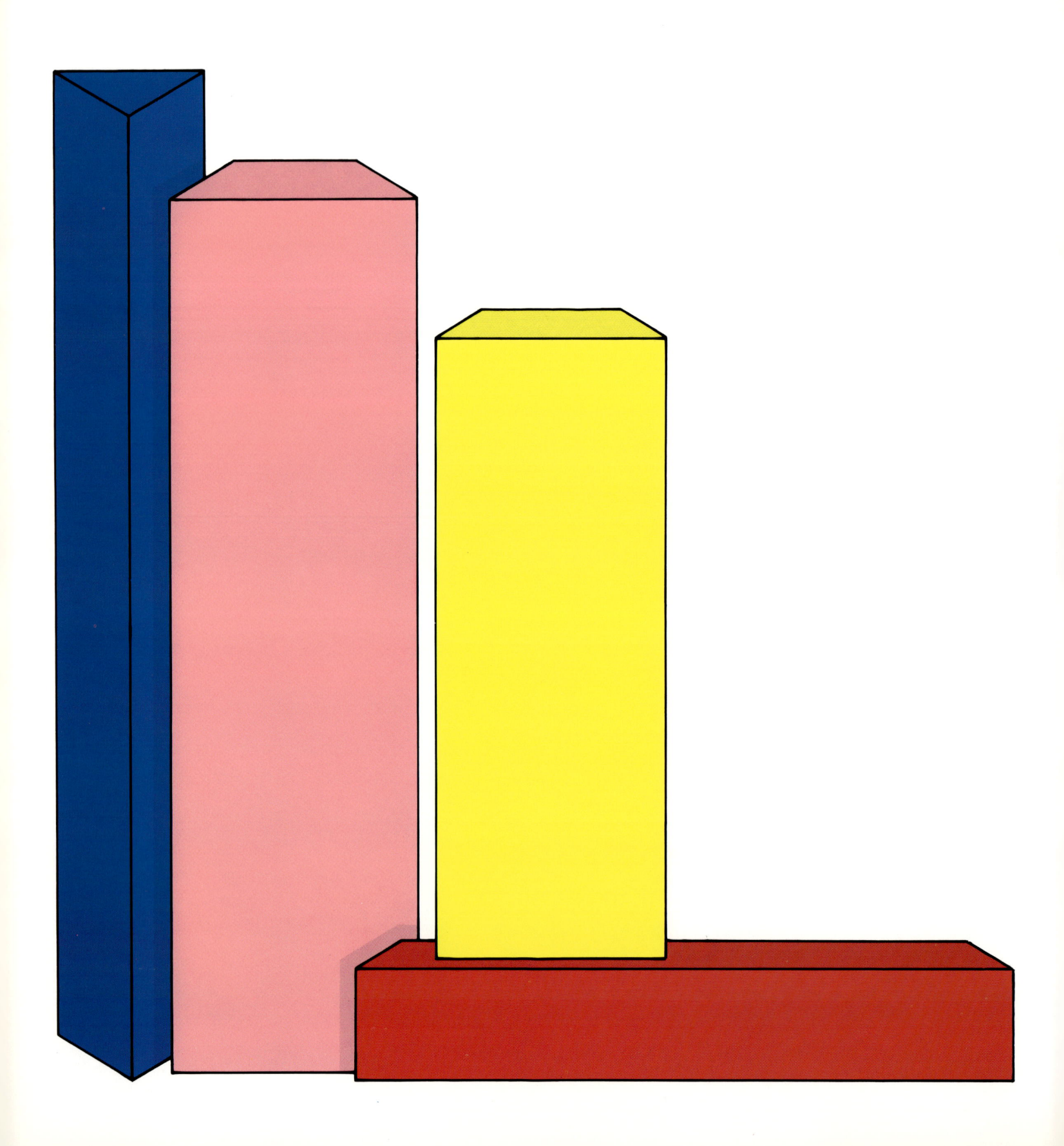

Only in recent years has the importance of packaging begun to be broadly recognized for its impact upon the long-term success of a brand. In fact, packaging may be the single most important media packaged goods have in which to introduce a product, compete in a volatile marketplace and sustain a competitive advantage.

Packaging is increasingly establishing its position as a cornerstone in the marketing mix and the exhibits on the following pages illustrate the commitment to excellence by the packaging team...designer, marketer, production and sales professional.

As the sophistication of marketers has increased, so has that of the designer. The packages exhibited on the following pages not only satisfy important distribution and merchandising requirements, they each clearly illustrate that a strong selling vehicle can be executed with finesse and style. It is interesting to note that those professionals designing packaging all over this globe share an uncommon commitment to excellence.

It is also interesting to consider that this publication goes beyond being a portfolio of attractive packaging. The pages represent a chronicle of design, technology and lifestyle of the world in which we live.

It is no small task to select one package which can be said, with confidence, to be the best of the best. Those which have been identified as PDC Gold Award Winners were chosen from hundreds of strong contenders. Obviously, the arts and science business, worldwide, is alive and well.

Paul R. Sensbach, Ph.D.
Director, Packaging & Creative Services
Marketing Dept., Kraft Inc.

182 Hardware

Entry:	Firewood & Forestry Packaging
Category:	Automotive/Hardware
Award:	Winner 1982
Design Firm:	Gregorry Fossella Associates

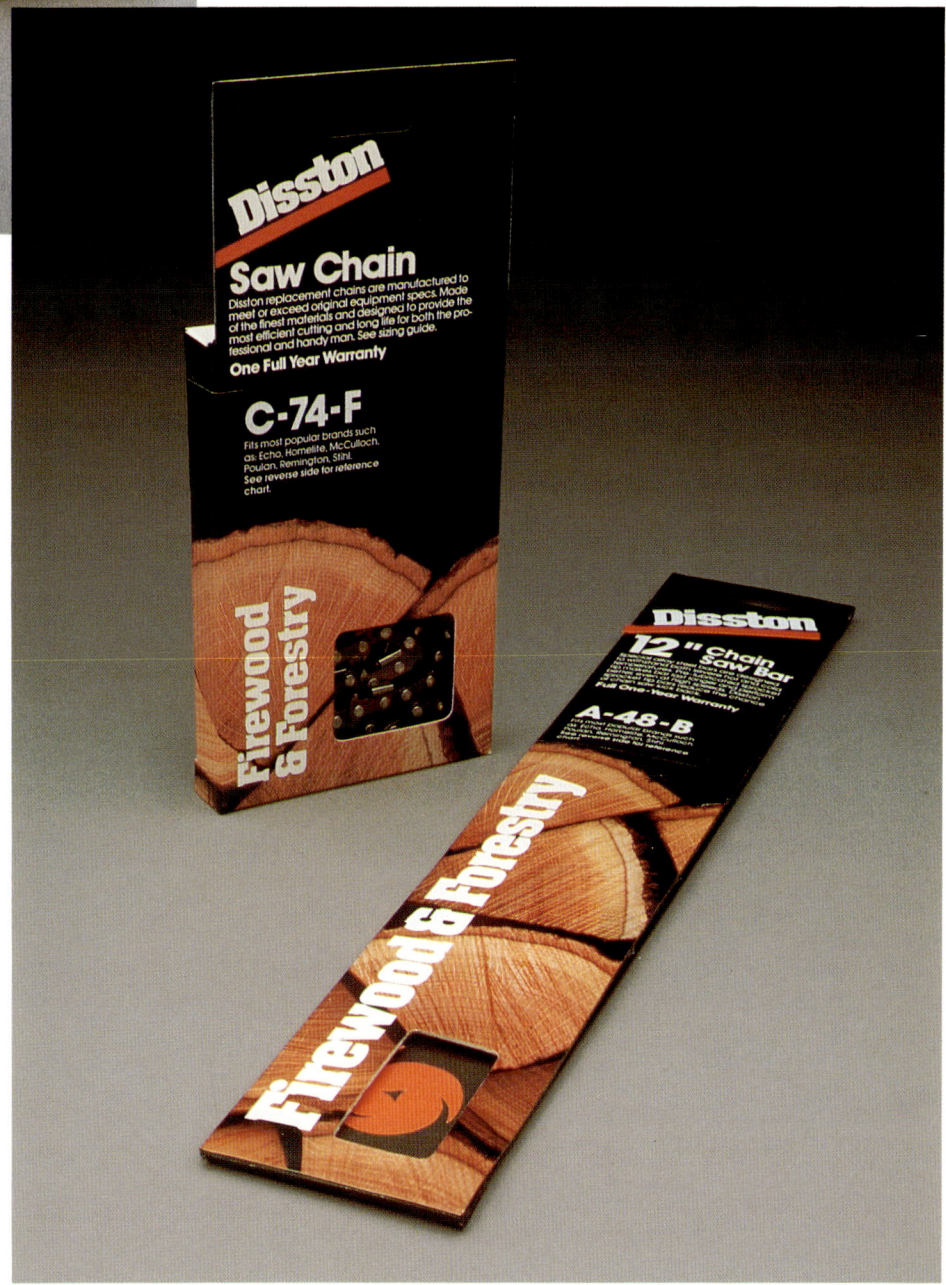

Entry	SKIL Power Tools
Category:	Automotive
Award:	Finalist 1985
Division:	USA
Design Firm:	Moonink Communications
Creative Director:	John Downs
Chief Designer:	John Downs
Client Firm:	Skil Corporation

Entry	Durabeam
Category:	Automotive
Award:	Finalist 1985
Division:	USA
Design Firm:	Macey Noyes Associates
Creative Director:	Androus Noyes, Kenneth Macey
Chief Designer:	Androus Noyes, Kenneth Macey
Client Firm:	Duracell U.S.A.

Entry	Ice Melt
Category:	Automotive/Garden/Hardware
Award:	Finalist 1987
Division:	USA
Design Firm:	Graphic Matrix Group, Inc.
Creative Director:	Robin Jastrzemski
Chief Designer:	Robin Jastrzemski
Client Firm:	LaRoche Industries Incorporated

186 Hardware

Entry	Applewood Herb Garden
Category:	Specialty
Award:	Winner 1987
Division:	USA
Design Firm:	TAB Graphics Design
Creative Director:	Tanis Bula
Chief Designer:	TAB Graphics Design/Illustrator—Darlene Emerick
Client Firm:	Applewood Seed Company

Entry	Herb Plant Kits
Category:	Miscellaneous
Award:	Finalist 1987
Division:	International
Design Firm:	Advertising Agency Tomoe
Creative Director:	Hisae Koyanagi
Chief Designer:	Hisashi Shimizu
Client Firm:	Japan Tobacco Inc.

Hardware 187

Entry	Petit Farm Plant Kit
Category:	Miscellaneous
Award:	Winner 1987
Division:	International
Design Firm:	Advertising Agency Tomoe
Creative Director:	Hisae Koyanagi
Chief Designer:	Hisashi Shimizu
Client Firm:	Japan Tobacco Inc.

Entry	Garden Gear
Category:	Automotive/Garden/Hardware
Award:	Finalist 1987
Division:	USA
Design Firm:	Van Noy Design Group
Creative Director:	Jim Van Noy
Chief Designer:	Lauren Harger
Client Firm:	McGuire-Nicholas Manufacturing Co.

188 Hardware

Entry	Probe
Category:	Automotive/Garden/Hardware
Award:	Finalist 1987
Division:	USA
Design Firm:	Peterson & Blyth Associates
Creative Director:	Richard Shear
Chief Designer:	Richard Shear
Client Firm:	Sandoz Crop Protection Corporation

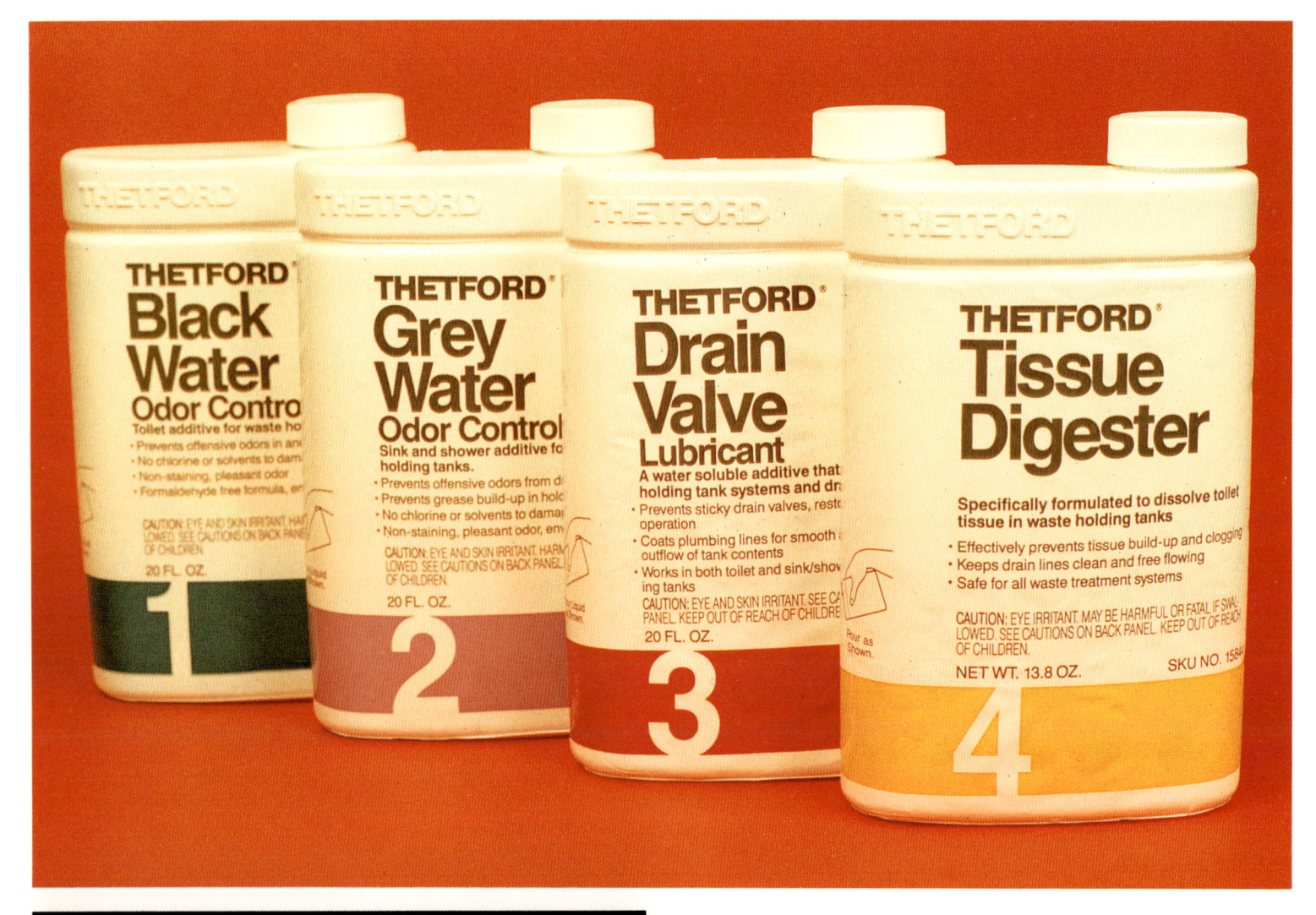

Hardware 189

Entry	Aqua-Chem (Pool Chemicals)
Category:	Automotive
Award:	Finalist 1985
Division:	USA
Design Firm:	Harte Yamashita & Forest
Creative Director:	Tets Yamashita
Chief Designer:	John Baker
Client Firm:	Georgia-Pacific Corporation

Entry	Preco Unleash
Category:	Pet Products
Award:	Finalist 1987
Division:	USA
Design Firm:	Murrie White Drummond Lienhart & Associates
Chief Designer:	Jayce Dougall
Client Firm:	Preco, Inc.

Hardware

Entry	Wagner Brake Fluid
Category:	Automotive/Garden/Hardware
Award:	Finalist 1987
Division:	USA
Design Firm:	Gerstman+Meyers
Creative Director:	Juan Concepcion
Chief Designer:	Rafael Feliciano
✤✤Design Supervision:	Richard Gerstman
Client Firm:	Wagner Division of Cooper Industries

Entry	Cosmo Motor Oil
Category:	Other
Award:	Finalist 1987
Division:	International
Design Firm:	Hakuhodo Inc.
Creative Director:	Tokihiko Kimata
Chief Designer:	Daizaburo Murakami
Client Firm:	Cosmo Oil Co., Ltd.

Entry:	AFT & LDO Automotive Oils
Category:	Automotive
Award:	Finalist 1985
Division:	USA
Design Firm:	Selame Design
Creative Director:	Joseph Selame
Chief Designer:	Joseph Selame
Client Firm:	Amoco Oil Company

Hardware

Hardware 193

Entry	Pro Fuel/Filter
Category:	Automotive
Award:	Winner 1981
Design Firm:	Dickens Design Group
Creative Director:	Robert Lee Dickens
Chief Designer:	Albert Yochim
Client Firm:	CR Industries

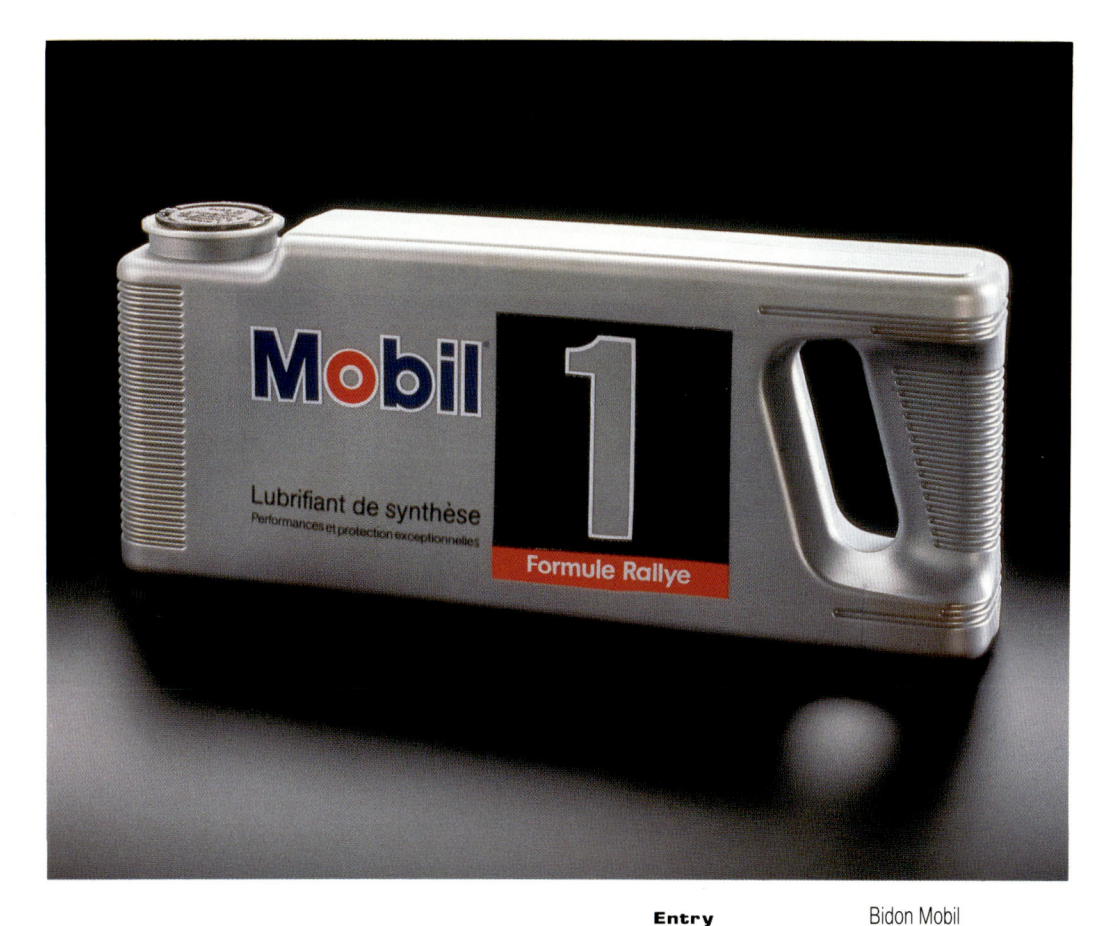

Entry	Bidon Mobil
Category:	Other
Award:	Finalist 1987
Division:	International
Design Firm:	Beautiful Design House
Creative Director:	Francis Metzler
Chief Designer:	Francis Metzler, Georges Lormelle
Client Firm:	Mobil Oil France

Entry:	Torque Starter Battery
Category:	Aut/Gdn/Hard
Award:	Finalist 1987
Division:	USA
Design Firm:	Spectrum Boston

194 Hardware

Entry: Norton Automotive Sandpaper & Accessories
Category: Automotive/Garden/Hardware
Award: Finalist 1987
Division: USA
Design Firm: Gamble And Bradshaw Design
Creative Director: Chet Makoski
Chief Designer: Jim Aresco, Judy Bochinski, Bill Phillips
Client Firm: Norton Company

Entry: Altec Lansing Speakers
Category: Automotive/Garden/Hardware
Award: Finalist 1987
Division: USA
Design Firm: Ronald Emmerling Design Inc.
Chief Designer: Ronald Emmerling
Client Firm: Altec Lansing

Hardware 195

Entry	Algreco Paints
Category:	Automotive
Award:	Finalist 1985
Division:	USA
Design Firm:	Gerstman+Meyers Inc.
Design Supervision:	Richard Gerstman
Creative Director:	Juan Concepcion
Chief Designer:	Joe Lombardo
Client Firm:	Borden Inc. International

196 Hardware

Entry	Bolt Paper Towel
Category:	Household
Award:	Finalist 1987
Division:	USA
Design Firm:	Lister Butler Inc.
Creative Director:	John Lister
Chief Designer:	Jerry Meyer
Client Firm:	James River Corporation

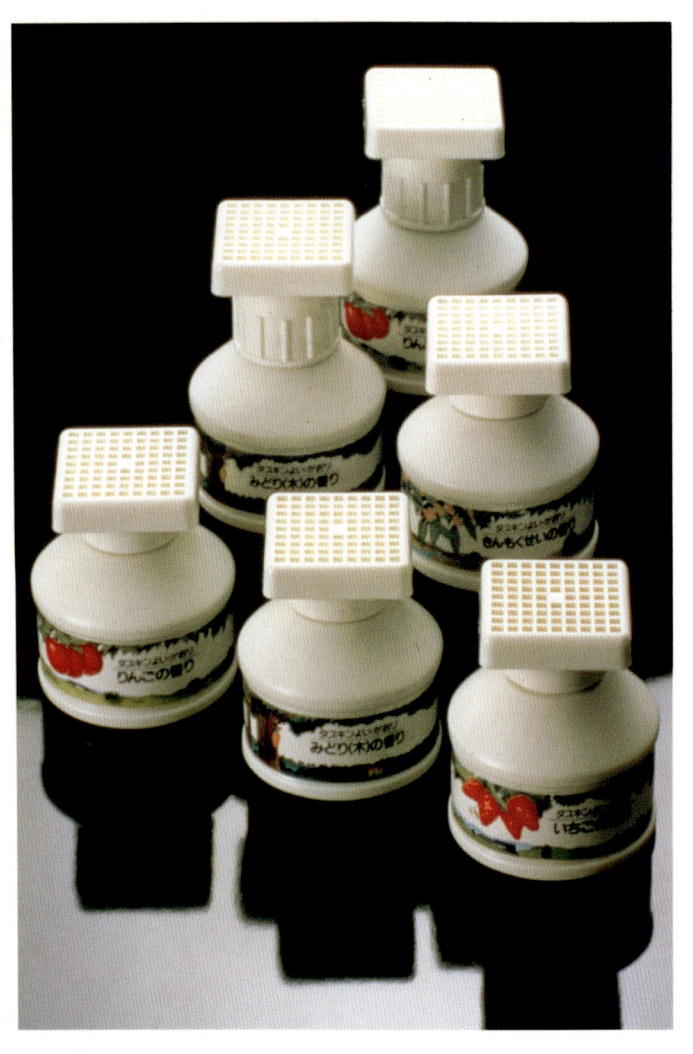

Entry:	Duskin Yoi Kaori Room Deodorizers
Category:	Household
Award:	Winner 1982
Design Firm:	Wakutsu Design Office Inc.

Entry:	Accu-u-sol Sprayer
Category:	Over-The-Counter
Award:	Finalist 1985
Division:	USA
Design Firm:	Precision Valve Corporation
Creative Director:	Robert H. Abplanalp
Chief Designer:	Virgil Naku
Client Firm:	Precision Valve Corporation

Hardware 197

INTRODUCING: ACC-U-SOL SPRAYER

NEW! FIRST TRIGGER SPRAY SYSTEM FOR PRESSURIZED PRODUCTS.

Now — a spray system to trigger the imagination of marketers! Opens a whole new world of aerosol ideas...heightens sales appeal and profits of current products. Contact us today for samples and the full story on how we can put ACC-U-SOL to work for you.

Continuous spray, no pumping • Low cost • Tamper-proof • Finger-tip control • Totally directional • Leak-proof

PRECISION
INNOVATIONS WORK!

PRECISION VALVE CORPORATION World Headquarters: Yonkers, NY 10702 USA
Manufacturing facilities throughout the world.

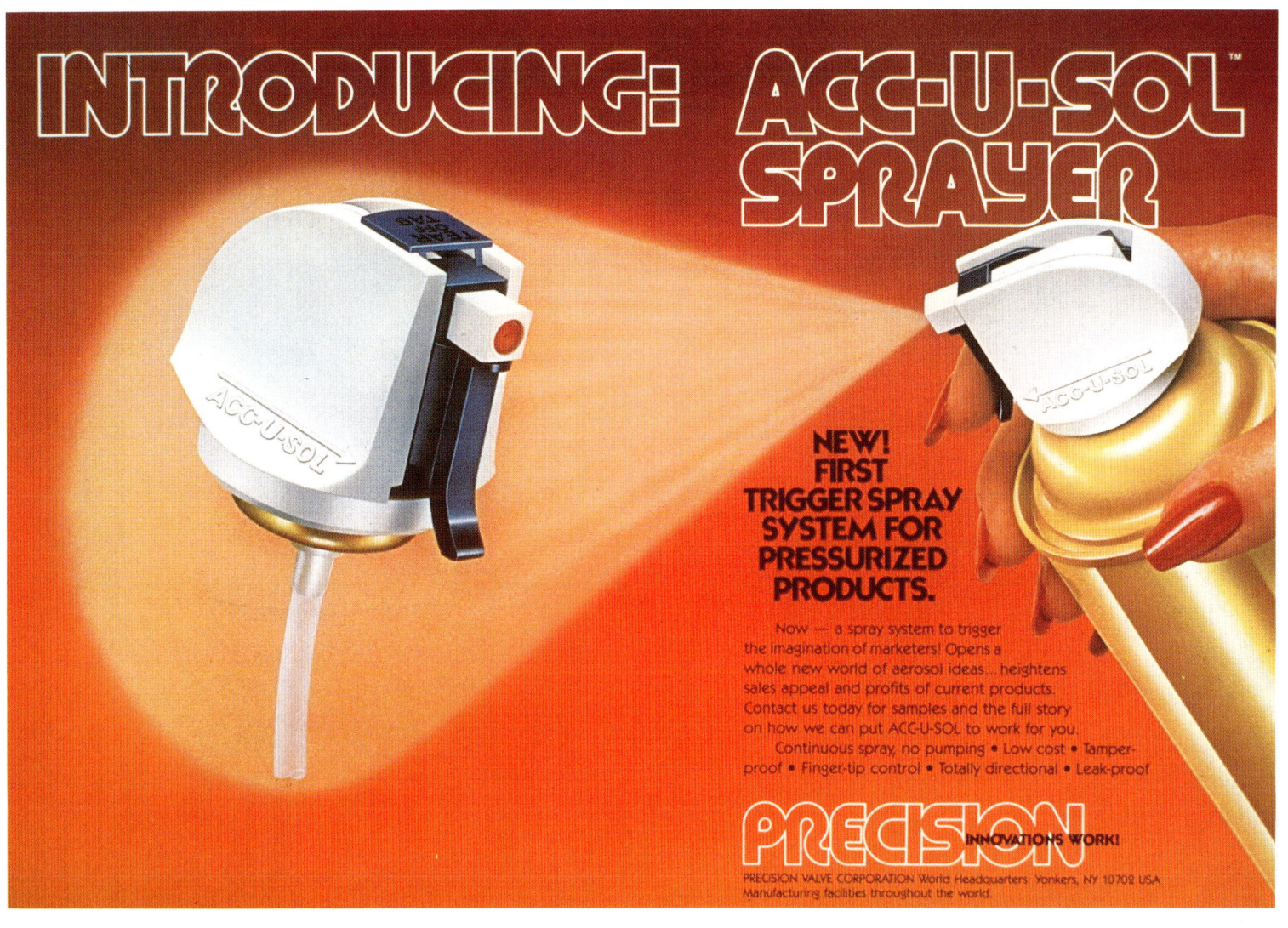

PRECISION
INNOVATIONS WORK!

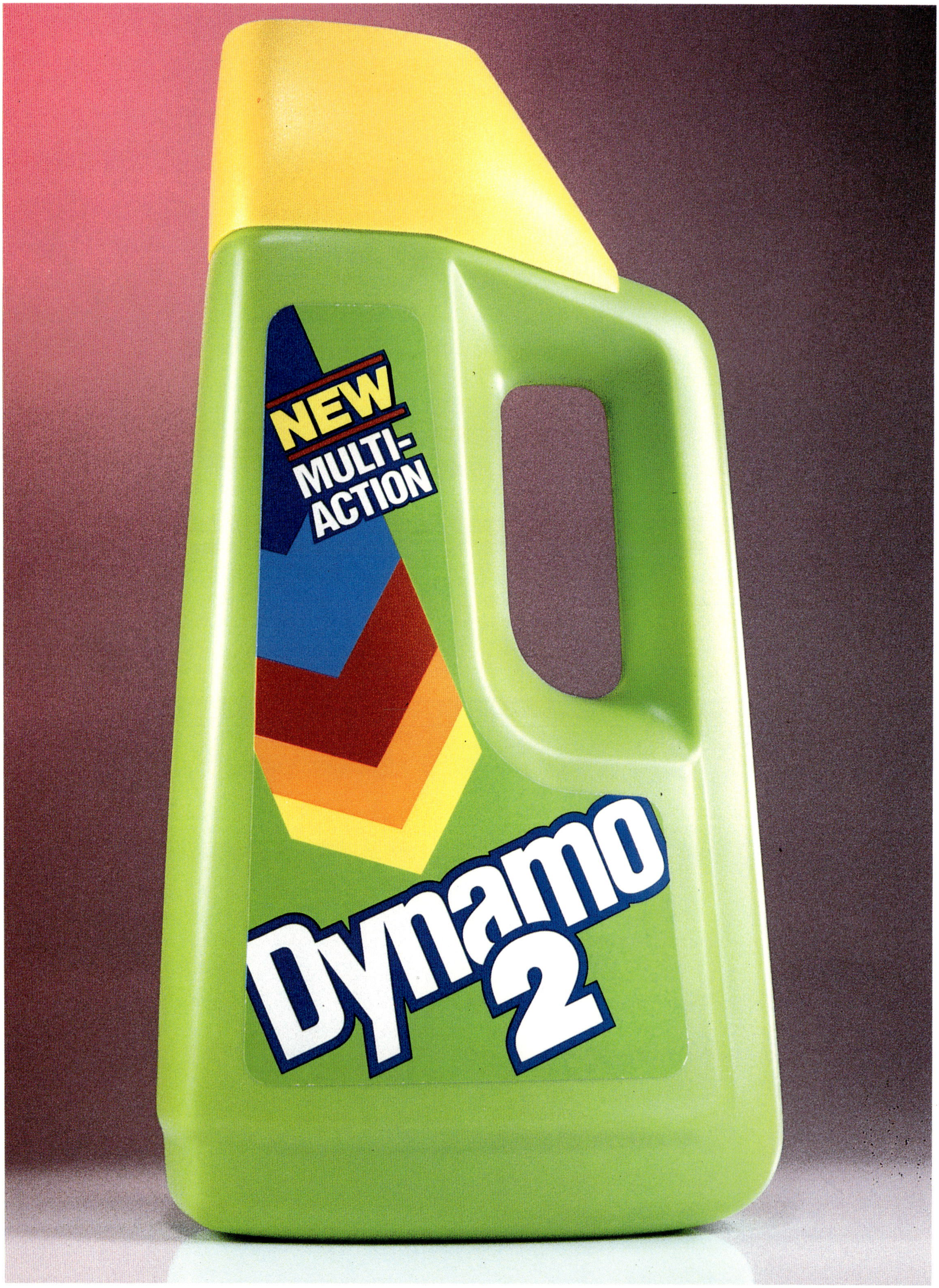
NEW
MULTI-
ACTION
Dynamo
2

Hardware 199

Entry: Dynamo 2
Category: Household
Award: Finalist 1985
Division: USA
Design Firm: Colgate Palmolive

Entry: Ajax All-Purpose Cleaner
Category: Household
Award: Finalist 1985
Division: USA
Design Firm: Colgate Palmolive

Entry Yes
Category: Household
Award: Finalist 1985
Division: USA
Design Firm: Luth+Katz
Creative Director: Alvin Katz
Chief Designer: Erdal Akdag
Client Firm: Texize

200 Hardware

Entry	Asda Private Label Washing Powders
Category:	Miscellaneous
Award:	Finalist 1987
Division:	International
Design Firm:	Lloyd Northover Limited
Creative Director:	John Lloyd
Chief Designer:	Simon Ingleton
Client Firm:	Asda Stores

Entry	Twice As Fresh Air Freshener
Category:	Household
Award:	Finalist 1987
Division:	USA
Design Firm:	Sidjakov Berman Gomez & Partners
Creative Director:	Nicolas Sidjakov, Jerry Berman
Chief Designer:	Barbara Vick
Client Firm:	Clorox Company

Hardware 201

202 Hardware

Entry	Fine Wood Floor Care Products
Category:	Automotive/Garden/Hardware
Award:	Finalist 1987
Division:	USA
Design Firm:	Source/Inc.
Creative Director:	Charles Cybul
Chief Designer:	Dale Joneson
Client Firm:	S.C. Johnson & Son, Inc.

Hardware 203

Entry	Murphy Oil Soap
Category:	Household
Award:	Finalist 1987
Division:	USA
Design Firm:	Gerstman+Meyers
Creative Director:	Juan Concepcion
❊❊Structural Designer:	Ric Hirst
❊❊Design Supervision:	Richard Gerstman
Client Firm:	The Murphy-Phoenix Company

Entry:	Carpet Care
Category:	Household
Award:	Finalist 1987
Division:	USA
Design Firm:	Spectrum Boston

CHAPTER 7
Retail and Softgoods/ Department Store Promotion

USA & Canada Division

The 1987 PDC Gold Awards are not representative of the package design one might encounter on a typical shopping trip, but rather the best of the breed.

In many cases, the design elements are blended to suggest qualities, not merely indicate package contents. In the face of an increasingly complex web of package identity regulations, one can detect a minimalist trend, that is, the nurturing and preservation of white space. This is particularly evident in the Purina O.N.E. package, the Henri Nestle and the Harvest Fresh Herb Garden as well as the Zaca Mesa wine bottles.

It's also interesting to note that packagers in a variety of markets are finding the design impact of bags to be very appealing. In the current rundown, we have General Foods' Maxwell House Private Collection, Nestle's Henri Nestle Chocolates, the Harvest Fresh Herb Garden as well as Purina. The dog food and herb garden might be considered traditional uses of bags and there certainly has been a use of bags for coffee for some time, but Nestle's use of them for chocolates and General Foods use of them for an upscale line is inventive.

International Division

The International Division of the PDC Gold Awards for 1987 is once again dominated by the Japanese entries which continue to show an inventive flair and a respect for quality, fidelity and fine detail. While these international packages catch the eye of many U.S. packagers, one has to realize the economics of producing such packages make them little more than idea stimulators in the U.S. market.

Ben Miyares
Executive Editor
Food & Drug Packaging

Entry	Gilmore's Retail Gift Packaging
Category:	Dept. Store
Award:	Winner 1981
Design Firm:	Container Corporation of America Design & Market Research Laboratory
Creative Director:	Raymond Peterson
Chief Designer:	Marvin Steck
Client Firm:	Gilmore's Department Stores

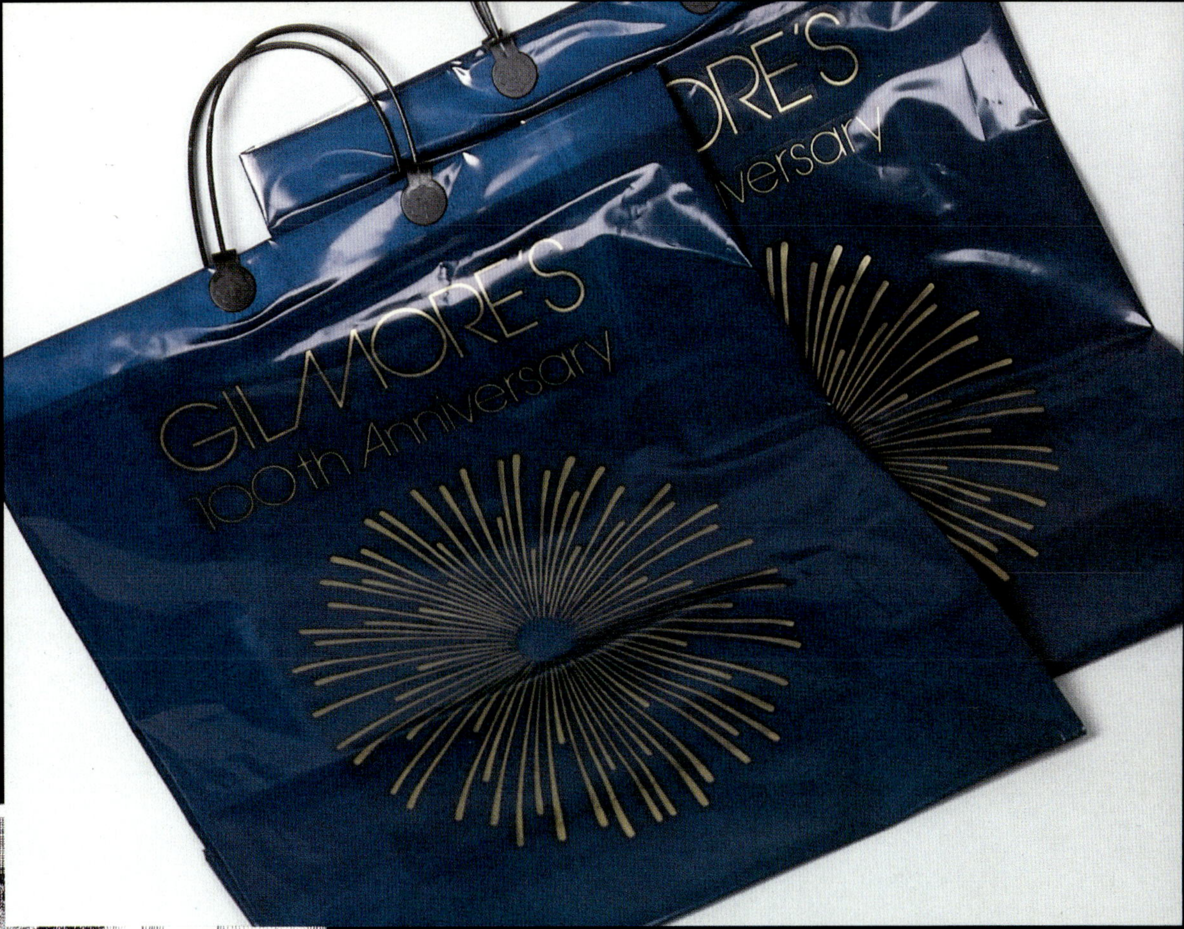
GILMORE'S
100 th Anniversary
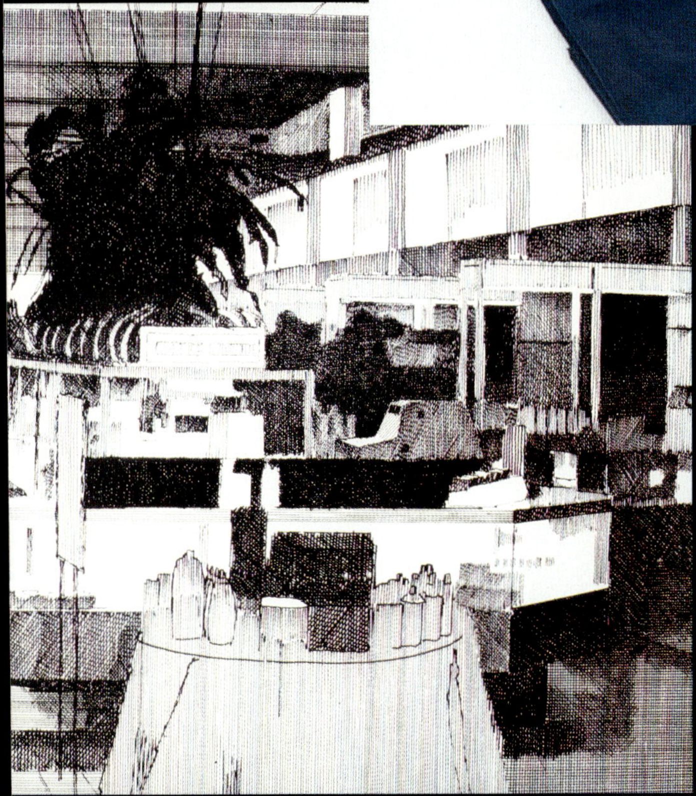

208 Retail & Softgoods/Department Store Promotion

Entry	Designer Salami
Category:	Promotional Packaging
Award:	Winner 1981
Design Firm:	Primo Angeli Graphics
Creative Director:	Primo Angeli
Chief Designer:	Primo Angeli, Mark Jones, Ray Honda
Client Firm:	P.G. Molinari & Sons

Retail & Softgoods/Department Store Promotion 209

Entry	Abercrombie & Fitch Shopping Bags
Category:	Department Store
Award:	Winner 1980
Design Firm:	GNU Group
Creative Director:	Richard Burns/James Reed
Chief Designer:	James Reed
Client Firm:	Abercrombie & Fitch

Entry	Oyster Bar Gift Packaging
Category:	Promo/Gift
Award:	Winner 1980
Design Firm:	Hans Flink Design Inc.
Creative Director:	Hans D. Flink
Chief Designer:	Hans D. Flink
Client Firm:	Grand Central Oyster Bar & Restaurant

210 Retail & Softgoods/Department Store Promotion

Retail & Softgoods/Department Store Promotion 211

Entry	Warners Undergarment
Category:	Displays
Award:	Winner 1982
Design Firm:	Joel Bronz Design Inc.
Creative Director:	Joel Bronz
Chief Designer:	Frank Hom
Client Firm:	Warner's

212 Retail & Softgoods/Department Store Promotion

Entry: Park Hyatt Guest Candy Box
Category: Promotional
Award: Winner 1982
Design Firm: Strandell Design Incorporated
Chief Designer: Don Strandell
Client Firm: Park Hyatt Hotel Chicago

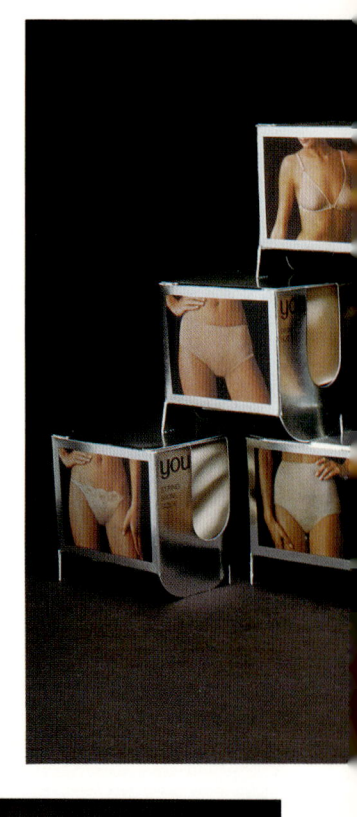

Entry: Totes Boot Line
Category: Retail
Award: Finalist 1985
Division: USA
Design Firm: Libby Perszyk Kathman Inc.
Creative Director: Tina Fritsche
Chief Designer: Liz Kathman Grubow
Client Firm: Totes, Inc.

Entry: Doral Saturnia International Spa
Category: Specialty
Award: Finalist
Division: USA
Design Firm: Cato Desgrippes Beauchant Gobe
Creative Director: Marc Gobe
Chief Designer: Marc Gobe, Gerben Hooykaas
Client Firm: Carol Management

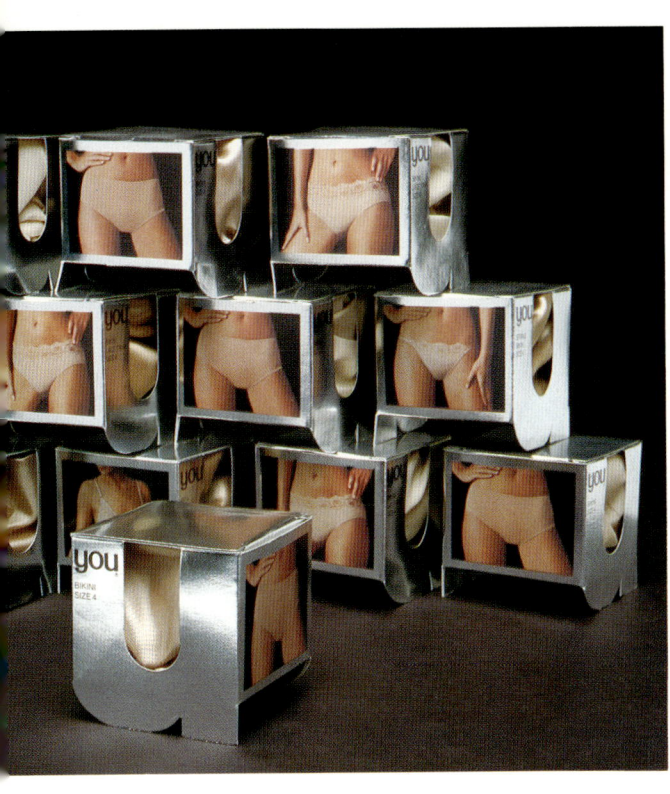

Retail & Softgoods/Department Store Promotion 213

Entry	"You"
Category:	Soft Goods
Award:	Winner 1981
Design Firm:	NW Ayer
Creative Director:	Van Templeton
Chief Designer:	Dante Evangelista
Client Firm:	Formfit Rogers

Entry	Boots Men's Gifts
Category:	Specialty
Award:	Finalist 1987
Division:	International
Design Firm:	Trickett & Webb
Chief Designer:	Lynn Trickett, Brian Webb & Andrew Thomas
Client Firm:	The Boots Company

Entry	Espringe Lingerie Series
Category:	Specialty
Award:	Finalist 1987
Division:	International
Design Firm:	Work Fabric Inc.
Creative Director:	Makiko Shiga
Chief Designer:	Yasuhiro Wakutsu
Client Firm:	Duskin Company Ltd.

214 Retail & Softgoods/Department Store Promotion

Retail & Softgoods/Department Store Promotion 215

Entry:	Loisdar
Category:	Department Store
Award:	Winner 1982
Design Firm:	Creative Five Company

CHAPTER 8
Office Supplies

**What Constitutes a Gold Award
To Be Selected by the Package
Design Council International**

Why have you entered this competition seeking a gold award? Prestige? Status? A statue on the desk? A plaque on the wall? New business? Undoubtedly, all of the above. But, above all...foremost...should be the real satisfaction of having produced a package of merit. A package that does its job well of selling the product, increasing market share and, generating profit.

 As a designer, the satisfaction comes from knowing that you have practiced your craft to the best of your ability; that you have used your imagination; that you have not resorted to a hackneyed answer to the problem nor created a cliche response to the market 'out there'. Your package should be contributing, just as the Gold Award itself, which intends to assist and set a guiding light to the young people in our profession and students who attain to be a part of it.

 The Package Design Council looks for those reasons when evaluating a Gold Award winner...so that our purpose as a professional organization also comes to fruition.

      May Bender, FPDC
      May Bender Design Associates

218 Office Supplies

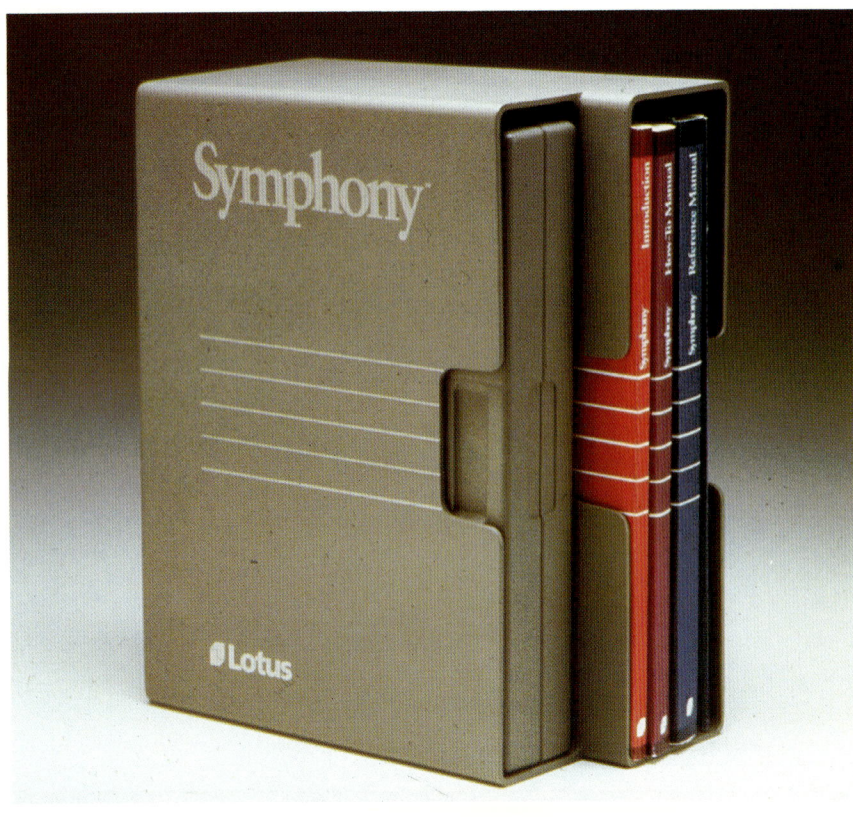

Entry:	Symphony Packaging
Category:	Office
Award:	Finalist 1985
Division:	USA
Design Firm:	Gregory Fossella Associates
Creative Director:	
Chief Designer:	
Client Firm:	

Office Supplies 219

Entry	Chase Global Disk Pack
Category:	Specialty
Award:	Finalist 1987
Division:	USA
Design Firm:	Pentagram Design
Creative Director:	Peter Harrison, Susan Hochbaum
Chief Designer:	Susan Hochbaum
Client Firm:	Chase Manhattan Bank

220 Office Supplies

Entry:	Datext Software
Category:	OFC
Award:	Finalist 1987
Division:	USA
Design Firm:	Gregory Fossella Design
Creative Director:	Michael Catacchio
Chief Designer:	Michael Catacchio
Client Firm:	Datext Inc.

Entry	Nukote International Packaging
Category:	Office
Award:	Winner 1987
Division:	USA
Design Firm:	Samata Associates
Creative Director:	Greg Samate
Chief Designer:	Jim Hardy
Client Firm:	Nukote International

Office Supplies 221

Entry	Data General Computer Supplies
Category:	Office
Award:	Finalist 1985
Division:	USA
Design Firm:	Hammond Design Associates
Creative Director:	Duane Hammond
Chief Designer:	David Roberts
Client Firm:	Data General

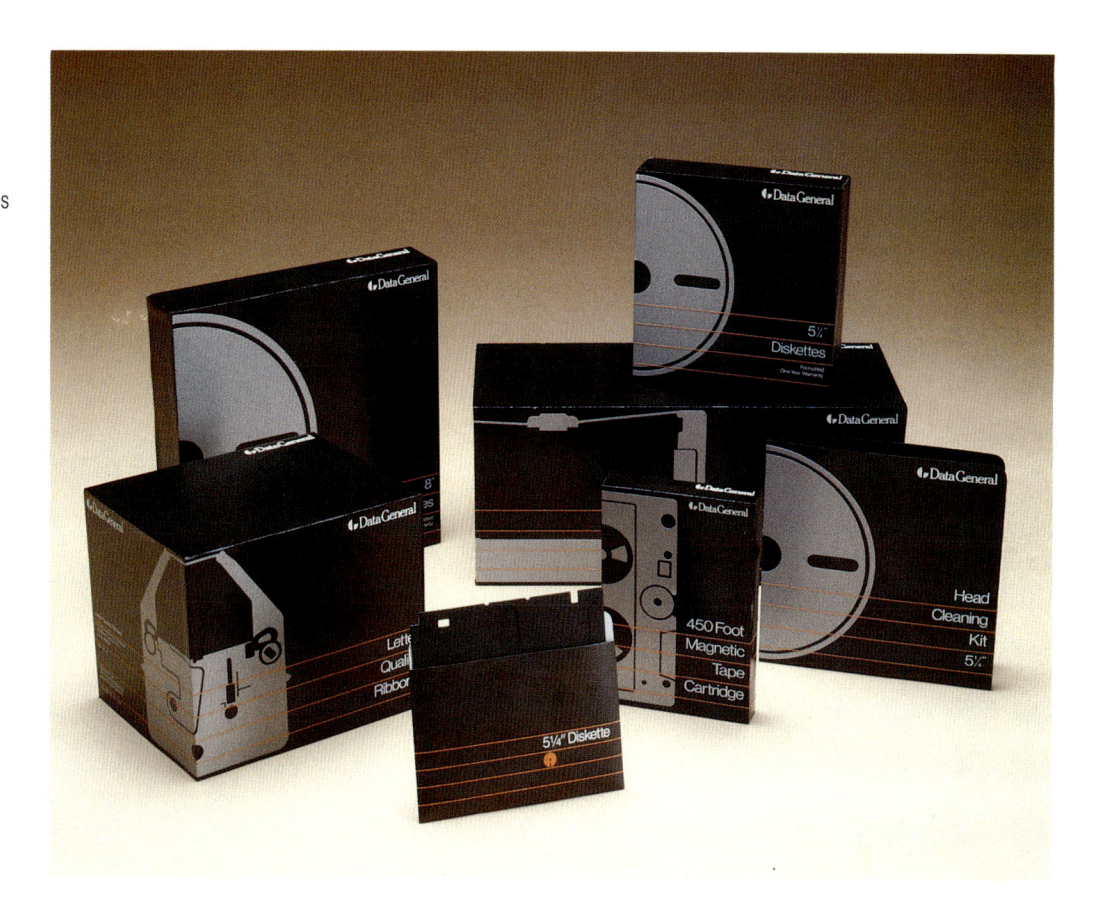

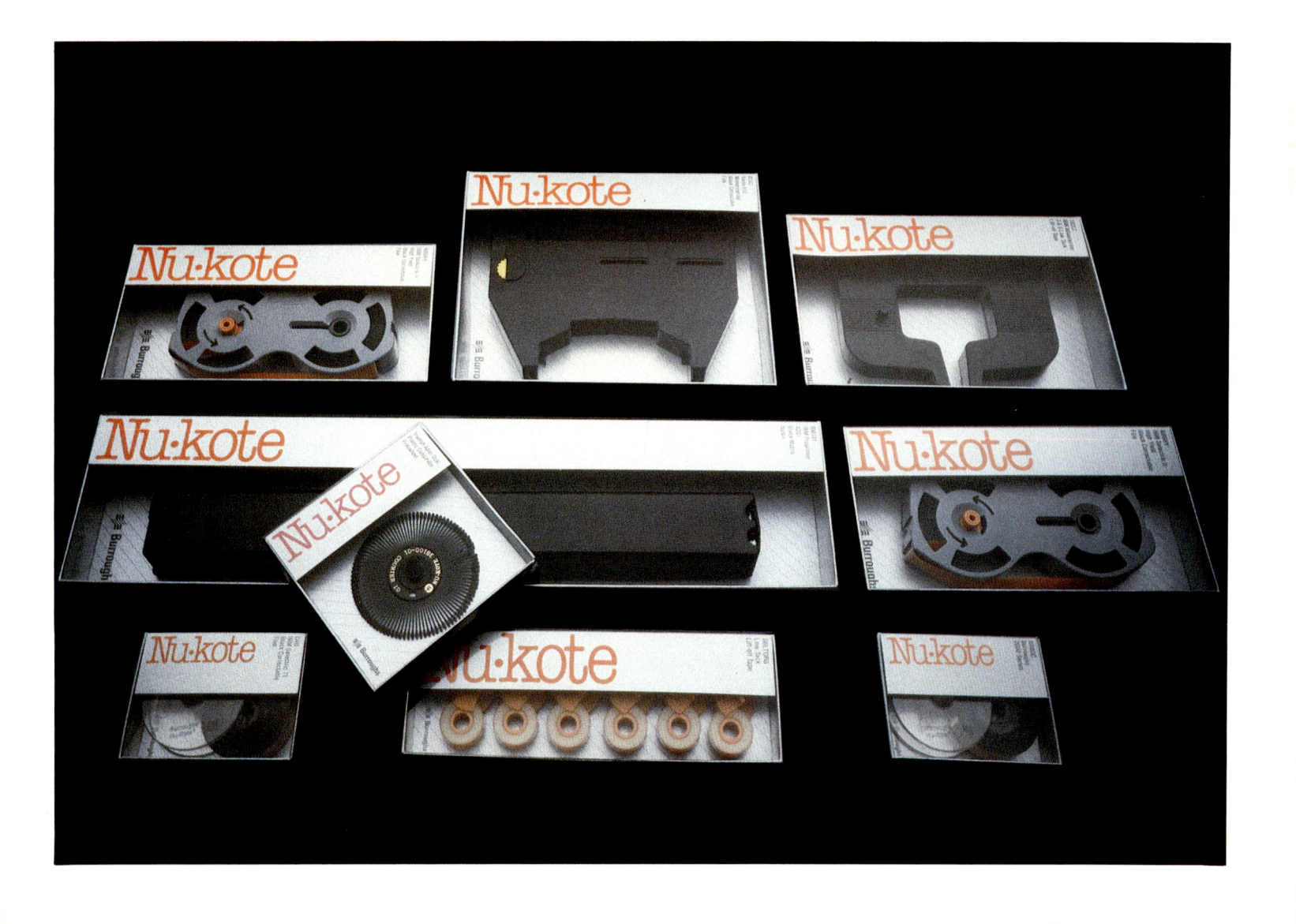

222 Office Supplies

Entry:	Digital Accessories
Category:	Office
Award:	Finalist 1985
Division:	USA
Design Firm:	Digital Equipment

Entry	Gold Hill
Category:	Miscellaneous
Award:	Finalist 1987
Division:	USA
Design Firm:	Selame Design
Creative Director:	Fred Golinko
Chief Designer:	Joseph Selame
Client Firm:	Goldhill Computers

Office Supplies 223

Entry	Up Front Software
Category:	Office
Award:	Finalist 1987
Division:	USA
Design Firm:	Whelan Design Office
Creative Director:	Richard J. Whelan
Chief Designer:	Jim Keller
Client Firm:	Dialcom, Inc.

Entry	IBMacBridge
Category:	Office
Award:	Finalist 1985
Division:	USA
Design Firm:	Lee Payne Associates, Inc.
Creative Director:	Martin J. Westley
Chief Designer:	Martin J. Westley
Client Firm:	Tangent Technologies

Office Supplies 225

Entry: Genicom
Category: Office
Award: Finalist 1987
Division: USA
Design Firm: Robert Hain Associates

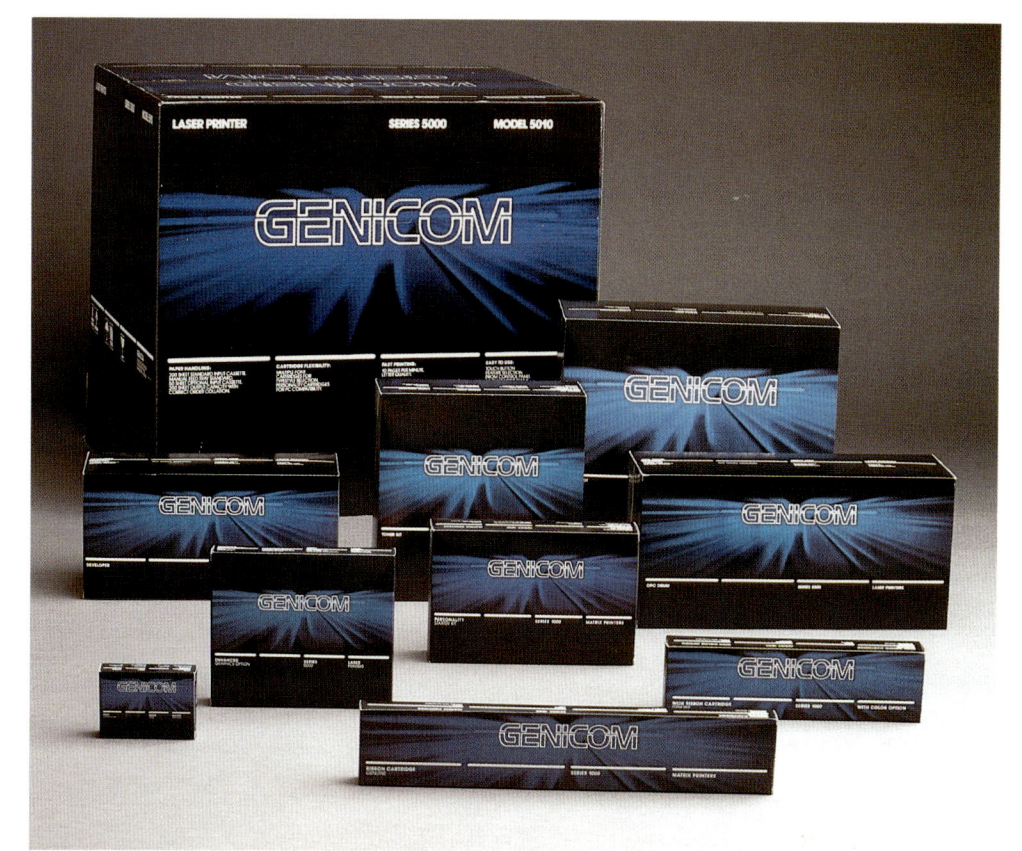

Entry: Okidata Microline
Category: Office
Award: Winner 1985
Division: USA
Design Firm: Robert Hain Assoc.

226　Office Supplies

Entry	Soft Strip System
Category:	Office Products
Award:	Finalist 1987
Division:	USA
Design Firm:	Si Friedman Associates, Inc.
Creative Director:	Si Friedman
Chief Designer:	Si Friedman
Client Firm:	Cauzin Systems, Inc.

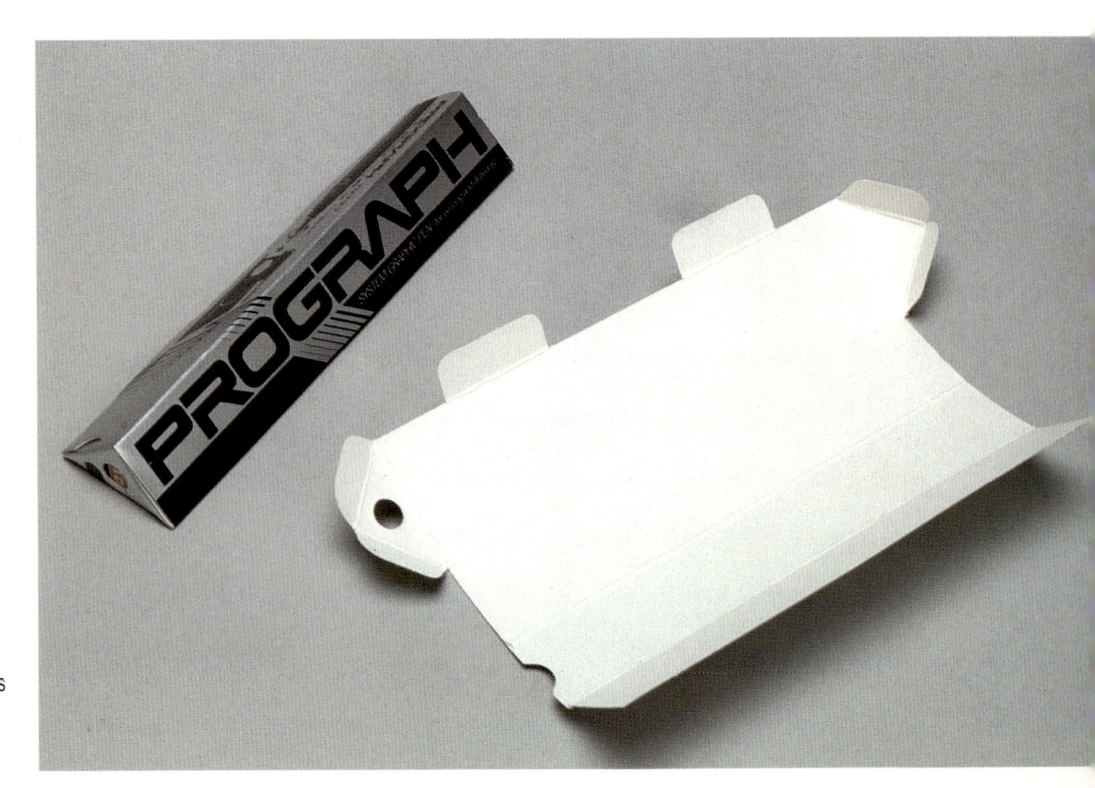

Entry:	Tombow Prograph Pens
Category:	Office Products
Award:	Winner 1982
Design Firm:	Zero Designs

Office Supplies 227

CHAPTER **9**

Corporate Identity

Successful packaging means different things to different people. To the technical and financial staff of a manufacturer, a successful package protects the product, is efficient on the packaging production line, easy to ship and costs little money. To the marketing man or the product manager, a successful package is one that executes the marketing strategy and creates an increase in sales. To the advertising executive, a successful package is one that creates a striking and memorable image that can be utilized in advertising and promotion. To corporate management, the successful package is one that helps increase profits.

Government people examine the package from the point of view of legal requirements. Health officials are concerned with nutritional information and health connotations where applicable.

Sanitation men and environmentalists look for a package that is easily disposable. And—to the designer, the successful package is one that does all these things and is aesthetically striking and pleasing.

Even though the package seems terribly important to each of these people representing different points of view and disciplines, it is virtually taken for granted by the consumer, unless it is annoying in some way. When the package just attracts the consumer to buy, communicates the proper information, is memorable enough to recognize for re-purchase, protects the product and is easy to open, reclose and dispose of, the consumer gives it little attention.

So, designing a successful package—one that the consumer will be attracted to, buy for first trial and easily recognize for re-purchase, is a formidable task.

Many products in the western world are so similar in quality that the package is the only thing that makes the difference. No matter how much advertising is used for a new product, at the point of sale it is usually the package that is the final convincing argument to buy.

Package designers are basically marketing people. They are practitioners of the art and science of consumer marketing. The art of selling useful everyday products to the average consumer—the massive population that buys and uses innumerable items for everyday needs and for leisure and enjoyment. The package designer specializes in the art of putting those everyday items into the most inexpensive and the most desirable packages from every point of view.

Package Designers Council is the organization that represents the professionals who strive to attain the highest level of accomplishment in their chosen profession.

Since the marketing person who is a package designer has usually been trained as an idealistic artist, the concept of encouraging talented young people in art schools to specialize in this demanding profession is appropriate.

The encouragement takes two forms. First, there is financial aid provided by the receipts of the PDC Gold Awards. Second, and probably more important, is the example of the highest levels of accomplishment set by the awards.

With over 30,000 different national, regional and private brands of package products on the supermarket shelves in the U.S. and more than 5,000 new items being introduced each year, choosing the few packages that are worthy of awards is difficult.

The Gold Awards mean different things to different people. Because a successful package must accomplish many, varied goals, the criteria for judging are hard to establish. Although there are deserving package designs that are not entered in the competition, the Gold Awards finalists represent the very highest in U.S. packaging aesthetics and quality.

Irv Koons, Executive V.P.
Siegel & Gale

Concours
Eurovision
de la Chanson
Bruxelles
Belgique

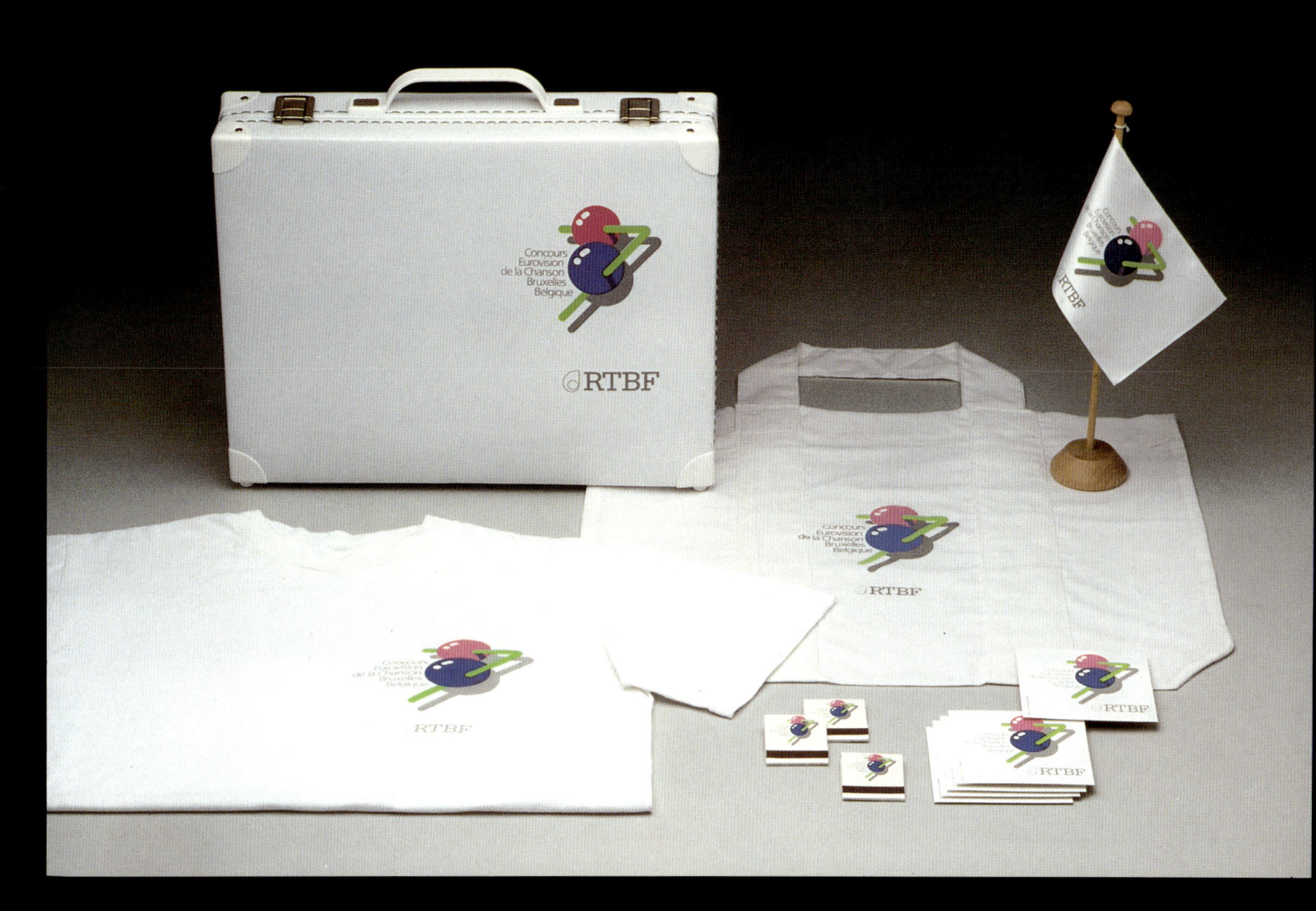

Corporate Identity 231

232 Corporate Identity

Entry	Certicare Merchandising Identity Program
Category:	Corporate Identity
Award:	Winner 1981
Design Firm:	Selame Design
Creative Director:	Joe Selame
Chief Designer:	Selame Design Group
Client Firm:	Amoco Oil Company

Corporate Identity 233

234 Corporate Identity

Entry	Dunlop
Category:	Corporate Identity
Award:	Finalist 1987
Division:	International
Design Firm:	Beautiful
Creative Director:	Francis Metzler
Chief Designer:	Francis Metzler, Andre Serrel
Client Firm:	Dunlop France

Corporate Identity 235

Entry	Anew
Category:	Corporate Identity
Award:	Finalist 1987
Division:	International
Design Firm:	Landor Associated
Creative Director:	Richard Kung, L. Creighton Dinsmore
Chief Designer:	L. Creighton Dinsmore
Client Firm:	Naturally Yours, Company

236　Corporate Identity

Entry:	Preferred Temporary Services
Category:	Corporate Image
Award:	Finalist 1987
Division:	USA
Design Firm:	Robert Hain Associates

Entry	Omni Hotels
Category:	Corporate Identity
Award:	Finalist 1987
Division:	USA
Design Firm:	Gerstman+Meyers
Creative Director:	Juan Concepcion
Chief Designer:	Sandy Meyers
❖❖Design	
Supervision:	Herbert M. Meyers
Client Firm:	The Omni Hotels

Corporate Identity 237

Entry	Sargento Cheese Company
Category:	Corporate Identity
Award:	Finalist 1987
Division:	USA
Design Firm:	Weidig Exhibits
Chief Designer:	Ron Horbinski
Client Firm:	Sargento Cheese Company, Inc.

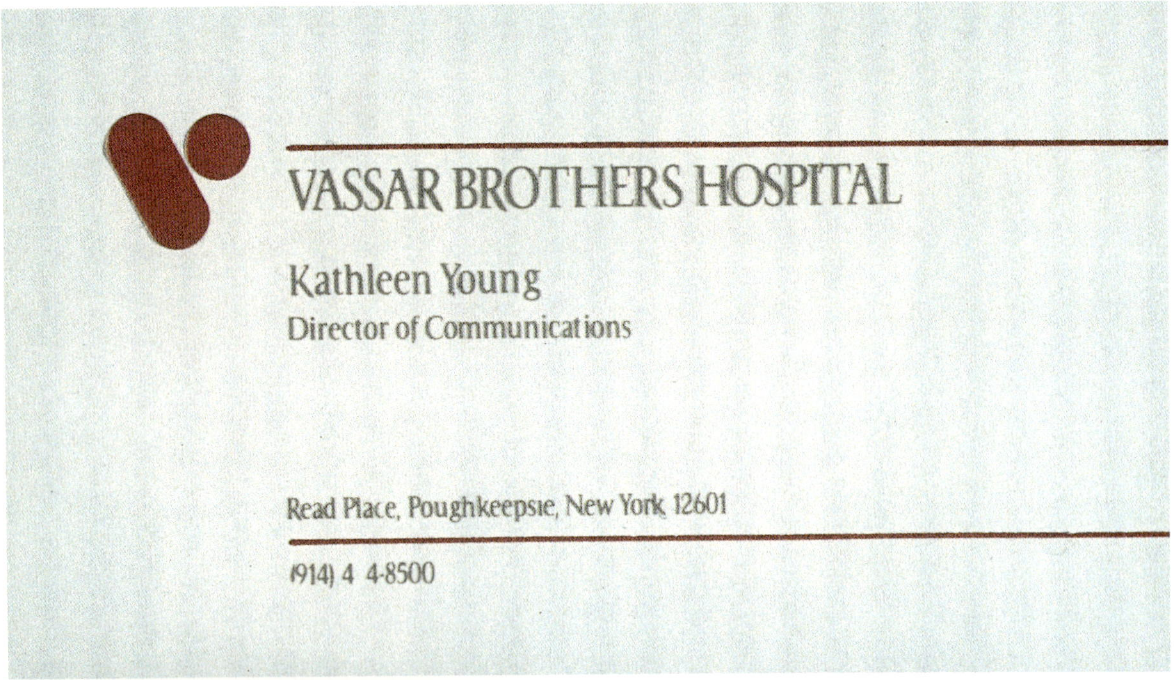

238 Corporate Identity

Entry: Vassar Brothers Hospital
Category: Corporate Identity
Award: Finalist 1987
Division: USA
Design Firm: Group Four Design

Corporate Identity 239

Entry	Wang Sattelite Sampler
Category:	Specialty
Award:	Finalist 1987
Division:	USA
Design Firm:	Innotech Corporation
Creative Director:	Jeff Cook/Lou Mennella
Chief Designer:	Dale Friedman/John Gusmano
Client Firm:	Wang Laboratories

240 Corporate Identity

Entry	Doral Saturnia International Spa
Category:	Corporate Identity
Award:	Finalist 1987
Division:	USA
Design Firm:	Cato Desgrippes Beauchant Gobe
Creative Director:	Marc Gobe
Chief Designer:	Marc Gobe, Gerben Hooykaas
Client Firm:	Carol Management

Corporate Identity 241

242 Corporate Identity

Entry	Liberty Savings Bank
Category:	Corporate Identity
Award:	Finalist 1987
Division:	USA
Design Firm:	Gerstman+Meyers
Creative Director:	Juan Concepcion
Chief Designer:	Karen Corell
Design Supervision:	Herbert M. Meyers
Client Firm:	The Liberty Savings Bank

Corporate Identity 243

Entry CTM Proposal Package
Category: Specialty
Award: Finalist
Division: USA
Design Firm: Florville Design & Analysis Inc.
Creative Director: Patrick Florville
Chief Designer: Joseph Bush
Client Firm: Citibank, N.A.

244 Corporate Identity

Entry	DHL
Category:	Corporate Identity
Award:	Finalist 1987
Division:	USA
Design Firm:	Primo Angeli Inc.
Creative Director:	Primo Angeli
Chief Designer:	Primo Angeli, Ray Honda, Eric Read
Client Firm:	DHL Worldwide Express

Corporate Identity 245

Entry:	Micropro
Category:	Corporate Identity
Award:	Winner 1982
Design Firm:	Charles Biondo Design Associates
Creative Director:	Charles Biondo
Client Firm:	MicroPro International Corp.

246 Corporate Identity

Corporate Identity 247

Entry	Sumdum
Category:	Corporate Identity
Award:	Finalist 1987
Division:	USA
Design Firm:	S&O Consultants Inc.
Creative Director:	David E. Canaan
Chief Designer:	Jacqueline Ghosin
Client Firm:	Arthur E. Hall, Integon Corporation

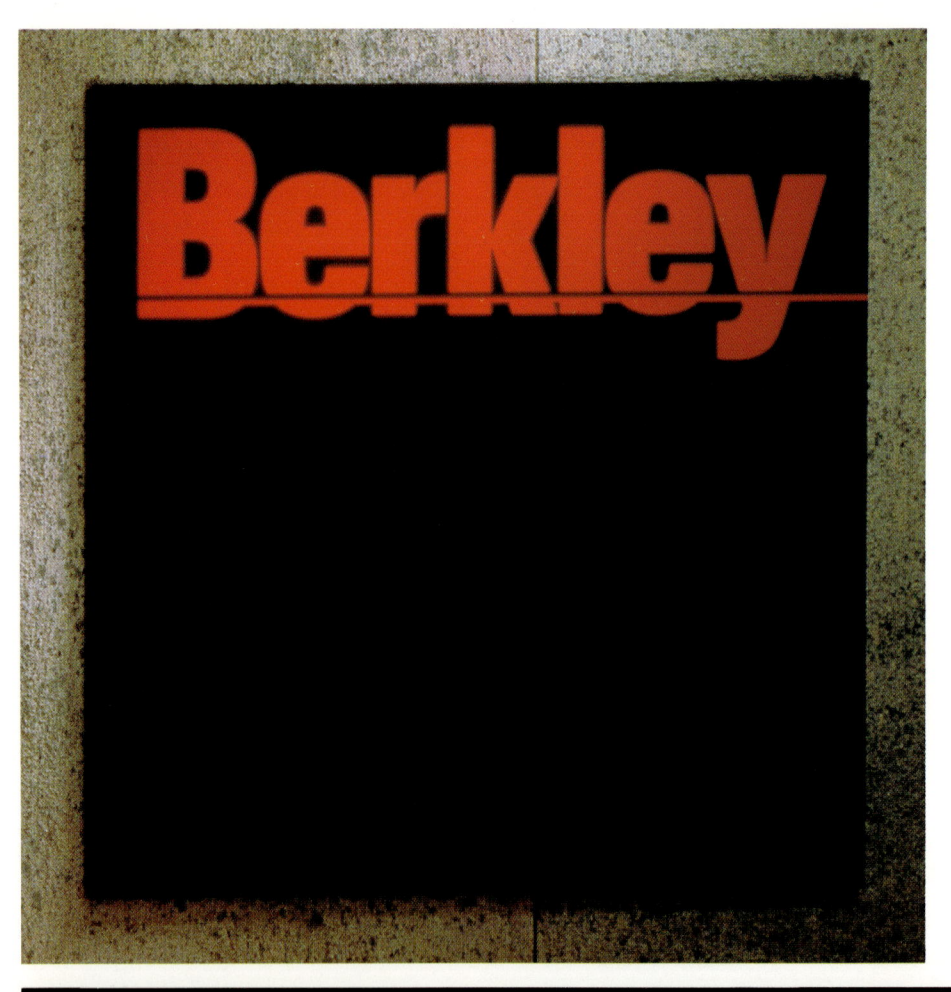

248 Corporate Identity

Corporate Identity 249

Entry	Berkley
Category:	Corporate Identity
Award:	Finalist 1987
Division:	USA
Design Firm:	Sidjakov Berman Gomez & Partners
Creative Director:	Nicolas Sidjakov, Jerry Berman
Chief Designer:	Courtney Reeser
Client Firm:	Berkley

250 Corporate Identity

INDEX

CREATIVE DIRECTORS

CHIEF DESIGNERS

CLIENTS

DESIGN FIRMS

OTHER CATEGORIES

Creative Directors

Abplanalp, Robert L. *197*
Adams, Dave *178*
Adams, Gaylord *66, 164*
Adams, Dave *102*
Angeli, Primo *88, 102, 104, 117, 125, 132, 208, 244*
Applebaum, Harvey *156*
Aoki, Shigeyoski *150*

Babcock & Schmid Associates, Inc. *54*
Bagby, Steven *146*
Behaeghen, Julien *230, 231*
Berman, Jerry *16, 30, 41, 48, 101, 107, 110, 113, 172, 174, 201, 249*
Berni Spinelli, Miller Brewing Co. *120*
Berni, Stephen L. (PDC) *44, 82, 137*
Berni, Stuart M. (FPDC) *106*
Biondo, Charles *245*
Blyth, John S. *20*
Bright, Keith *38, 149*
Broeders, Ad *165*
Bronz, Joel *106, 211*
Brubaker, Dennis *126, 134*
Bruce, George *173*
Bula, Tanis *186*
Burns, Richard *209*

Calo, Louise *113*
Canaan, David E. *247*
Catacchio, Michael *220*
Casey, Eugene J. Jr. *163*
Church, Stanley *21, 71*
Chwast, Seymour *59, 116, 161*
Coleman, Owen *32*
Colonna, Ralph *89, 94*
Concepcion, Juan *14, 26, 41, 42, 43, 45, 52, 59, 79, 103, 114, 125, 160, 170, 190, 194, 203, 236, 242*
Copp, Jeff *117*
Cook, Jeff *239*
Cybul, Charles *105, 115, 169, 202*

D'Addazio, Thomas *98*
Daigle, Charles *127*
Dentsu Inc. *96*
Desdoigts, Ouver *121, 140*
Dimino, Vincent *158*
Dinsmore, L. Creighton *235*
Downs, John L. *16, 50, 51, 184*

Emmerling, Ronald *179*
Ewry, Edwin *112*

Farrell, John *89*
Flink, Hans D. *209*
Flock, Don *60*
Florville, Patrick *175*
Fraternale, Marion *112*
Friedman, Michael *95*
Friedman, S. *133, 226*
Fritsche, Tina *212*

Gabel, Jim *49*
Gangi, Guy *82*
Gobe, Marc *134, 135, 136, 212, 240*
Golinko, Fred *222*
Grubow, Liz Kathman *126*

Hammond, Duane *221*
Harbauer, Jerry *57, 68*
Haruyo, Tahira *153*
Heming, Rowland S.W. *35, 66, 105*
Hochbaum, Susan *219*
Harrison, Peter *219*

Ingleton, Simon *69*
Inuzuka, Tatsumi *57*

Jastrzemski, Robin *185*
Juett, Dennis S. *147*
Jutsum, Alec *168*

Kaneko, Shuya *63*
Kaplan, Karin *128*
Kass, Warren A. *93*
Katz, Alvin *199*
Keller, Denis *60*
Kimata, Tokihiko *128, 191*
Klein, Larry *38*
Kondo, Fumio *33*
Koons, Irv *43, 170*
Koyanagi, Hisae *186, 187*
Kung, Richard *235*

Lafortezza, Michael *70, 101*
Leonard, Peggy *179*
Lewis, Mary *15, 25, 92, 96, 97, 111, 151*
Lipson, Alport, Glass & Assoc. *17, 22, 47, 51, 81, 83*
Lister, John *22, 23, 31, 32, 50, 61, 196*
Livolsi, Michael *147*
Lloyd, John *200*

Macaulay, D.F. *19, 34, 56, 63*
Maeda, Kazuki *72*
Makoski, Chet *173, 194*
Umebara, Makoto *15*
Mennella, Lou *239*
Marchi, Giancarlo Italo *30, 70, 74*
Markis, Beverly *35*
Matsui, Keizo *46, 73*
Metzler, Francis *37, 234*
Mittleman, Fred *100*
Morrill, Edward *127*
Musios, Timothy S. *145, 151, 171*

Noyes, Androus *185*

O'Connor, William J. *159*

Parcels, J. Roy *58*
Perszyk, Ray *40, 117*
Peterson & Blyth Associates, Inc. *75, 76*
Peterson, Raymond *45, 206*
Peterson, Ronald *129, 130, 131*
Porter, Allen *162, 163*

Reed, James *209*
Robertz, Henry John *132*
Rokudo, Mitsuro *48*

Salt, Thomas R. *139*
Samate, Greg *221*
Sandstrom, David *35, 127, 128, 168*
Sato, Makoto *90, 91*
Satoh, Taku *150*
Sawachi, Yasuyuki *143*
Scarlett, David *20, 65, 175*
Selame, Joe *232, 233*
Shear, Richard *188*
Shiga, Makiko *213*
Shiozawa, Yoshinori *90, 91, 99*
Shiozu, Hiroki *128*
Shiratani, Keizo *46, 99, 119*
Sidjakov, Nicolas *16, 30, 41, 48, 101, 107, 110, 113, 172, 174, 201, 249*
Stout, Kay *18, 146*
Sugiura, Shunsaku *143*

Tahiri, Haruyo *141*
Tajimi, Hajime *109*
Tazumi, Shiro *38, 56, 62, 67*
Templeton, Van *213*
Tipton, Judy *65*
Tree Trapanese *179*

Van Noy, Jim *187*

Wallace, Robert *109*
Werbin, Irv *98*
Westley, Martin J. *224*
Whelan, Richard J. *223*
Wineries Guild *95*

Yamashita, Tets *27, 41, 69, 189*
Yano, Kaneto *131, 140*
Yao, Takeo *36, 47, 73*
Yazawa, Shozo *62*
Yonetsu, Hisakichi *39, 77*
Young, Dick *147*

Chief Designers

Adams, Gaylord **60, 164**
Akdag, Erdal **199**
Akgulian, Mark **120**
Angeli, Primo **88, 102, 104, 117, 125, 208, 244**
Applebaum, Harvy **156**
Aresco, Jim **194**
Aron, Michael **59, 116, 161**

Babcock & Schmid Associates, Inc. **54**
Baker, John **189**
Barnhart, Kevin **61**
Barrett, Janice **40**
Bechtold, John (photographer) **156**
Berard, Jerome **111, 151**
Bergman, Mark **30**
Berni, Stuart M. (FPDC) **44, 82, 137**
Blyth, Jack **124**
Bochinski, Judy **194**
Bond, Thomas **110**
Botthoff, John **56, 63**
Broeders, Ad **165**
Bush, Joseph **243**

Cagney, William **132**
Cajoleas, Lotis **115**
Catacchio, Michael **220**
Cero, Vicki **104**
Chan, Phillis **21, 109**
Chwast, Seymour **59, 116, 161**
Claps, Louis **163**
Clark, Dale **179**
Collier, Michelle **94**
Colonna, Ralph **89**
Coop, Jeffery **57**
Corell, Karen **43, 125, 170, 242**
Cruanas, Bob **65**
Cybul, Charles **159**

D'Addazio, Adam **98**
DiFranza, Steve **128**
Dinsmore, L. Creighton **235**
Dolph, Bernard **105, 169**
Dougall, Jayce **189**
Downs, John L. **50, 184**

Ellert, Lisa **162**
Emmerling, Ronald **178, 194**
Evangelista, Dante **213**

Feliciano, Rafael **52, 190**
Florville, Patrick **175**
Foshuag, Jackie **110, 172, 174**
Friedman, Dale **239**
Friedman, J. **226**
Fuji, Kimiyo **57**
Flink, Hans D. **209**

Gabel, Jim **49**
Guaquelin, Benedicte **140**
Gee, Paul **43, 170**
Gerstman & Meyers Design Staff **114**
Ghosin, Jacqueline **247**
Gittings, Jane **146**
Gobe, Marc **134, 135, 136, 212, 240**
Goldner, Ken **133**
Goldsmith, Linda **173**
Grygienc, Marie V. **68**
Gusmano, John **239**

Hardy, Jim **221**
Harger, Lauren **187**
Haworth, Diane **124**
Hayashi, Chihiro **141**
Healy, Susan **27, 69**
Heming, Rowland, S.W. **66**
Hochbaum, Susan **219**

Hom, Frank **211**
Honda, Ray **88, 102, 104, 125, 132, 208, 244**
Hooper, Ward **127**
Horbinski, Ron **237**
Hooykaas, Gerben **212, 240**

Ingleton, Simon **200**

Jastrzemski, Robin **185**
Jones, Mark **88, 104, 117, 125, 132, 208**
Jones, Thomas **51**
Joneson, Dale **115, 202**
Juett, Dennis J. **147**

Kaneko, Takashi **143**
Kass, Warren A. **93**
Kathman Grubow, Liz **212**
Kawakami, Motomi **109**
Keller, Denis **139, 230, 231**
Keller, Jim **223**
Kim, Jung **106**

Lam Design Associates Design Dept. **70, 101**
Lee, Sokie **126**
Leonard, Peggy **179**
Levine, Catherine **23, 31**
Levine, Peter **134, 135, 136**
Lewis, Mary **15, 25, 92, 97**
Lipson, Alport, Glass & Assoc. **17, 22, 47, 51, 81, 83**
Lombardo, Joe **45, 195**
Lynch, Dick **61, 118**

Macey, Kenneth **185**
Maeda, Seiichi **72**
Maguire, Cynthia **89**
Mammina, Michael **171**
Marchi, Giancarlo Italo **70**
Marrington, Barbara **19**
Matsunaga, Shin **109**
Matthys, Daniele **105**
May, Gwen **171**
Merwin, Chris **22**
Metzler, Francis **37, 234**
Metz, John **58**
Meyer, Jerry **196**
Meyers, Sandy **236**
Miller, Judith **26**
Miller, William **163**
Mittleman, Fred **100**
Miyamoto, Yasushi **128**
Mizuno, Takuzo **47, 73**
Moonink Communications **16**

Nakayama, Eiko **63**
Naku, Virgil **197**
Nelson, Diane **112**
Nevins, James **16, 101, 113, 172**
Ng, Peggy **18**
Nikolits, Peter **113**

Ohama, Yoshitomo **145, 151**

Pearle, Carolyn **134**
Pennington, Alex **42**
Phillips, Bill **194**
Peterson & Blyth Design Group **20, 74, 76, 175**
Peterson, Ronald **129, 130, 131**
Porter, Allen **162, 163**

Racina, Amy **95**
Read, Eric **88, 244**
Reed, James **209**
Reeser, Courtney **174, 249**
Retger, Helen **50**
Riddell, Larry **14, 59, 79, 103, 160**
Riedell, Bill **98**

Roberts, David **221**
Rogers, Brian **35**
Rokudo, Mitsuru **48**
Rollinson, Mark **69**

Sandstrom, David **128**
Satoh, Taku **150**
Sands, Colin **165**
Sato, Makoto **90, 91**
Satoh, Taku **96**
Scarlett, David **20**
Schwendeman, David **127, 168**
Segar, Abe **32**
Selame Design Group **232, 233**
Selame, Joseph **102, 178, 222**
Sezzel, Andre **37, 234**
Shear, Richard **188**
Shelley, Barbara **65**
Sheintop, Linda **136**
Shelsey, Leonara M. **58**
Shimizu, Hisash **186, 187**
Shinyama, Sachiko **46, 73**
Shizatani, Keizo **46, 99**
Shiokawa, Minoru **143**
Shiozawa, Yoshinori **90, 91, 99**
Shiratani, Keizo **119**
Skelsey, Fiona **130**
Steck, Marvin **45, 206**
Sterling, William **34**
Sugunra, Shunsaku **150**
Sykes, Penny **159**
Szczygunski, Ed **52**

Tab Graphics Design **186**
Tabor, Robert **112**
Tuzumi, Shiro **38, 56, 62, 67**
Thomas, Andrew **213**
Togosawa, Tetsuo **150**
Toutain, Guillame **121**
Tree Trapanese **179**
Trickett, Lynn **130, 159, 165, 213**
Tucker, Leslie **71**
Tuzzesson, Ann **32**

Umebaza, Makoto **15**
Umezawa, Kazunori **145, 151**

Vallone, Kim Read **82**
Vantal, Erik **60**
Vick, Barbara **107, 174, 201**
Voll, Linda **44**

Wakutsu, Yasuhizo **131, 140, 213**
Watanabe, Hideo **169**
Waxman, Sabza **14, 59**
Webb, Brian **130, 159, 165, 213**
Weir, Shelly **125**
Wells, Jon **95**
Werbin, Irv **98**
Westley, Martin J. **224**
White, Jeff **53**
White, Thomas Q. **24**
Willoughby, Karen **106**
Wong, Hedy Lin **82, 159**
Wood, Neil **96**
Wood, Raymond **38, 149**

Yamada, Hiroko **153**
Yazawa, Shozo **62**
Yochim, Albert **193**
Yonetsu, Hisakichi **39, 77**

Zinander, Susan Bailey **40**

CLIENTS

Abercrombie & Fitch 209
Activision 172
Age International Inc. 98
Ajinomoto Co. Inc. 36, 47
Ajinomoto General Foods Inc. 109
Amoco Oil Company 232, 233
Allen Beverage Co., Ltd. 114
Anheuser-Bush, Inc. 118
Aoyagi Company 15
Applewood Seed Company 186
Arthur E. Hall, Integon Corp. 247
Asada Stores 69, 200
ASDA Stores Ltd. 92, 96
Associated Merchandising Corp. 161
Avon 136, 137
Avon Products Co., Ltd. 145
Avon Products Co., Inc. 151
Bankaku Sohonpo & Co. 39, 77
Beatrice—Hunt Wesson 20
Benjamin & Medwin Inc. 156
Berkley 174, 249
Beverage Products Ltd. 105
Biersdorf Inc. 133
Biscuits Delacre 66
Booth Seafood Company 22
Boots Co., The 25, 151, 213
Borden Inc. International 195
Brasserie de Solenes 121
Brighton Products 146
Bristol-Meyers 125
Brown Group-Etage (The) 134, 135
Buckingham Wine 100
Burger King Corporation 42
California Milk Advisory Board 107
Campbell's Fresh Business Unit 34
Campbell's Frozen Foods Business Unit 19
Campbell's Worldwide Export Co./D. Lazzaroni and C.S.P.A. 56, 63
Carme Inc. 125
Carnation Company 27
Carol Management 212, 240
Caurin Systems, Inc. 226
Celentano Brothers 26
Celestial Seasoning/Kraft, Inc. 47
Centre Brands 40
Charlotte Charles, Inc. 44
Chase Manhattan Bank 219
Chase & Sandborn Foods 113
Chateau Chevalier Winery 89
Chesebrough-Ponds 133
Chipwich 65
Christian Brothers 88
Citibank, N.A. 243
Clorox 201
Club des Createurs de Braute 140
Colgate—Palmolive Enterprises 134
Colgate—Palmolice/United Kingdom 126
Continental Baking Co. 31
Coopervision Opthalmic Products 146
Corning Glass Works 156, 157, 159, 161
CPC International 43
C R Industries 193
Culinar Foods Dry Biscuit Div 66
Data General 221
Datext, Inc. 220
Dialcom, Inc. 223
DHL Worldwide Express 244
Dole Foods 84
Dorma/Brinichaus 165
Dove International/Division of Mars, Inc. 81, 83
Dunlop France 234
DuPont 185
Duracell, USA 185
Duskin Company Ltd. 131, 140, 213
Eagle Snacks, Inc. 61
Eastman Kodak Company 178
Ektelon 173

Entenmann's 74
Farberware Inc. 160
Florville Design & Analysis 175
Foodways Nationals 18
Food Brands Groups Ltd. 111
Formfit Rogers 213
Francais Confectionery Co., Ltd. 73
Fujikko Co., Ltd. 46, 73
General Foods Corporation 21, 26, 28, 109
General Mills 65
Georgia-Pacific Corporation 189
Gillette Canada, Inc. 127
Gillette Co. (The) 127, 129
Gillette Company (The)/Personal Care Division 128
Gilmore's Department Stores 206
Gloria Ferrer Sonoma Champagne Caves 94
Goldhill Computers 222
Goten Yatsuhashi Co., Ltd. 62
Gourmet Resources 59
Grand Central Oyster Bar & Restaurant 209
Great Atlantic and Pacific Tea Company (A&P) 44
Guild Wineries 95
Hal Riney and Partners (For Blitz Weinhard Brewing Co.) 117
HBO/Cannon Video 179
Hershey Food Corporation 54
Heublein 101
Hills Brothers Coffee 110
Hirakawa Fugetsuda Co., Ltd. 56
H.J. Heinz Company 45
International Marketing Concepts, Inc. 163
Isetan Department Store, Tokyo 63
Jacobs Coffee Company 112
James B. Beam Distilling Co. 115
James River Corporation 196
Japan Foucher 62
Japan Tobacco Inc. 186, 187
John Frieda 130
Jovan, Inc. 132
Kanebo Home Products Sales, Ltd. 128
Keebler Company, The 57
Keystone Camera Corporation 178
Kentucky Fried Chicken Japan Ltd. 33
Kitamura Co. Ltd. 38
Kraft Inc. Dairy Group 79
Kroger Co., The 17, 49
Kobayashi Kose Co., Ltd. 141, 153
LaRoche Industries Incorp 185
Leaf, Inc. 68
Le Clic Products Inc. 179
Libby Perszyk Kathman Inc. 58
Liberty Savings Bank (The) 242
Lizianne Blue Plate Foods 43
Lightning Bug, Ltd. 162
Los Angeles Olympic Organizing Committee (LAOCC) 38
Madonoume Shuzo Co., Ltd. 90
Majani Candies 70
Matchbox Toys (USA) Ltd 169
Max Factor K.K. 150
Maxwell House Division of General Foods Corporation 113
MCA Advertising for Stroh Brewery 116
McCain Frima 35
McGuire-Nicholas Manufacturing Co. 187
Meitan Honpo Co., Ltd. 46
Merrell Dow Pharmaceuticals, Inc. 147
Metropolitan Opera Guild 58
Micropro International Corp. 245
Miller Brewing Company 120
Milton Bradley Company 168
Miyozakura Jyozo Co., Ltd. 99
Monogram Models, Inc. 171
Monoprix 37
Morinaga & Company Limited 57
Mrs. Paul's Kitchens 22

Murphy-Phoenix Company, The 203
Nannini Panettone 30
Naturally Yours, Company 235
Natural Nectar Corporation 69
Nestle Foods Corporation 61, 70, 106
Nikka Whiskey Distilling Co., Ltd. 96
Nishikiaji 48
Niwa Co., Ltd. 67
Noodles By Leonardo Inc. 45
North American Philips Corp. 163
Norton Company 194
Nukote International 221
Ocean Spray Cranberries 85
Omni Hotels, The 236
Paddington Corporation, The 93, 98
Panosh Place 170
Park Hyatt Hotel Chicago 212
Perugina Maitre Patissier 74
P.G. Molinari & Sons 208
Phillips International B.V. 165
Pillsbury Center 35
Pillsbury Co., The 41, 71
Plumrose 14
Pop Rocks Inc. 60
Precision Valve Corp. 197
Preco, Inc. 189
Procter & Gamble GMBH 126, 139
Pure Ltd./Hawaiin Water Partners 115
Quaker Oats Co., The 51, 82, 105
Radio-Television Belge De La Communaute Francaise 230, 231
Ralston Purina Company 16, 50, 51, 52, 53
R.H. Macy & Co., Inc. 112, 158
Richardson-Vicks 130, 131
Rich Product Corporation 55
R.J. Reynolds Foods, Inc. 20
Rutherford Estate Winery/Inglenook-Napa Valley 89
Sandoz Crop Protection Corp. 188
Sargento Cheese Company, Inc. 24, 237
S.C. Johnson & Son Inc. 124, 129, 202
Shaklee Corporation 132
Shibata Partnership 91
Shiseido Co., Ltd. 143, 150
Siku of America/Sieper Werke GmBh 170
Skil Corporation 184
Smith Food Group S.A. 60
Souhonke Surugaya 72
Spong PLC 159
Sunflo Juice 102
Sundor Brands, Inc. 106
Takara Shuzo Co., Ltd. 99, 119
Tangent Technologies 224
Teledyne Waterpik 164
Texize 199
Thomas J. Lipton Inc. 23, 32, 42, 103, 104
Totes, Inc. 212
Treesweet Products Co. 102, 104
Trimensa, Inc. 149
Tsuji Hoten Co., Ltd. 90, 91
United Brands Company 82
United Rum Merchants (U.K.) 97
Universal Foods 48
Van Dutch 107
Venture Foods 64
Victoria's Secret 136
Wm. T. Thompson Company 147
Wagner Division of Cooper Industries 190
Wang Laboratories 239
Warner's 211
Weetabix Company 16
Welch Foods, Inc. 101
Whitestone Products 137
Wilson Sporting Goods 173
Worthington Foods 40
Zaca Mesa Winery 95

DESIGN FIRMS

Ad Broeders Graphic Design B.V. **165**
Advertising Agency Tomoe **186, 187**
Ajinomoto Co. **36**
Albion Cosmetics Co./Design Room **144, 153**
The Applebaum Company **156**
Avon **153**
Avon Products Co., Ltd. **145, 151, 171**

Babcock and Schmid Associates, Inc. **54**
Bagby Design Incorporated **146**
Beautiful **234**
Beautiful Design House **37, 121, 140**
The Berni Company **44, 82, 106, 137**
Bright & Assoc. **38, 149**

Campbell Soup Co. Design Center **19, 34, 56, 63**
Cato Desgrippes Beauchant Gobe **134, 135, 136, 212, 240**
C'Bon Cosmetics Co., Ltd. **108**
Charles Biondo Design Associates **18, 26, 28, 64, 84, 156, 157, 161, 164, 245**
Coleman, Lipuma, Segal, & Morrill, Inc. **32, 127, 129**
Colgate-Palmolive **198, 199**
Colgate Corporate Design Group **126, 134**
Colonna, Farrelli: Design Associates **89, 94**
Container Corporation of America Design & Market Research Laboratory **45, 206**

Dennis S. Juett & Associates, Inc. **147**
Designed To Print **179**
Design Board/Behaeghel & Partners S.A. **60, 139, 230, 231**
Design North **120**
Design Office Mail Box **15**
Digital Equipment **222**
Dixon & Parcels Associates, Inc. **58**

Florville Design & Analysis Inc. **175, 243**

Gamble & Bradshaw Design **173, 195**
Gaylord Adams/Don Flock & Associates **60, 164**
General Foods Corporate Design Center **113**
Gerstman & Meyers **14, 26, 41, 42, 43, 45, 52, 59, 79, 103, 114, 125, 160, 170, 190, 195, 203, 236, 242**
Giancarlo Italo Marchi **30, 70, 74**
GK Graphics Associates **63**
GNU Group **209**
Graphic & Package Design Studio Siro **46, 99, 119**
Gregory Fossella Associates **125, 183, 218, 220**
Group Four Design **238**

Hakuhodo Inc. **128, 191**
Hammond Design Associates **221**
Hans D. Flink **209**
Harte, Yamashita, & Forest **27, 40, 69, 189**

Innotech Corporation **239**
Irv Koons Associates, Inc. **43, 170**

James River Corporation **78**
JC Penney In-House Creative Services **160**
Joel Bronz Design Inc. **106, 211**
Jon Wells Graphic Design (carton graphics) **95**
Toppan Printing Company (structural design)
Joss Design Group **171**

Kass Communications **93**
Keizo Matsui & Associates **46, 73**
King Casey, Inc. **163**
Kobayashi Kose Co., Ltd. **141, 153**
Kollberg/Johnson Associates, Inc. **85, 104, 133**

Landor Associates **18, 55, 115, 146, 147, 235**
Lee Payne Associates, Inc. **224**
Lewis Moberly **15, 25, 92, 96, 97, 111, 151**
Libby Perszyk Kathman Inc. **40, 49, 58, 117, 126, 212**

Lipson Alport Glass & Associates **17, 22, 47, 51, 81, 83**
Lister Butler Inc. **22, 23, 31, 32, 50, 61, 196**
Lloyd Northover Limited **69, 200**
Lock/Pettersen **139**
Luth & Katz **199**

Macey Noyes Assoc. **185**
Meada Design Associates **72**
McCann-Erickson Hakuhodo **33**
Mittleman/Robinson Design Associates **100**
Lam Design Associates, Inc. **70, 101**
Lewis Moberly **151**
Moonink Communications **16, 50, 51, 184**
Morinaga & Company Limited Design Room **57**
Murrie, White, Drummond & Lienhart **24, 44, 53, 189**

Nason Design Associates, Inc. **35, 127, 128, 168**
Nestle Foods Corporation **80**
NW Ayer **213**

Obata Design, Inc. **61, 118**

Par Art Room **108**
Par Art Co. Ltd. **38, 56, 62, 67**
Parker Brothers **169**
Pentagram Design **219**
Peterson & Blyth Associates, Inc. **20, 65, 74, 76, 124, 129, 130, 131, 175, 188**
Pineapple Design **35, 66, 105**
Porter/Matjasich & Associates **162, 163**
Precision Valve Corporation **197**
Primo Angeli, Inc. **88, 102, 104, 117, 125, 132, 208, 244**
The Pushpin Group Inc. **59, 116, 161**

Rich Co., Ltd. **90, 99**
R.H. Macy & Co., Inc. Corporate Graphics **112, 158**
Robert Hain Associates **176, 177, 225, 235**
Robertz Webb & Co. **132**
Rokudo Design Office **48**
Ronald Emmerling Design Inc. **178, 179, 194**

Samata Associates **221**
SC Johnson & Son, Inc. **124**
Schering Plough **138**
S & O Consultants Inc. **112, 247**
Selame Design **102, 178, 222, 232, 233**
Shincom Co., Ltd. **75**
Shin Matsunaga Design Office, Inc. & Motomi Kawakami Design Room **109**
Shiseido **150**
Shiseido Co., Ltd./N.Y. **143**
Si Friedman Associates, Inc. **133, 226**
Sidjakov Berman Gomez & Partners **16, 30, 41, 48, 101, 107, 113, 172, 174, 201, 249**
Source/Inc. **82, 105, 115, 159, 169, 202**
Spectrum Boston **203**
Strandell Design Inc. **212**

TAB Graphics Design **186**
Taku Satoh Design Office **96, 150**
The Creative Source, Inc. **98**
The Thomas Pigeon Design Group Ltd. **53, 66**
Tipton & Maglione **65**
Trickette & Webb **130, 159, 165, 213**

Van Noy Design Group **187**

Wakutsu Design Office Inc. **196**
Wallace Church Associates **21, 71, 109**
Wallner, Harbauer, Bruce & Assoc. **57, 68, 173**
Weidig Exhibits **237**
Whelan Design Office **223**
Werbin Associates **98**
Work Fabric, Inc. **131, 141, 213**

Yao Design Institute Inc. **36, 47, 73**
Yazawa Design Office **62**
Yonetsu Design House **39**

Zero Designs **76, 226**

Other Categories

Biondo, Charles *18, 26, 28, 64, 84, 156, 157, 161*
Davnys, Joella *109*
Emerick, Darlene *186*
Gerstman, Richard *26, 41, 43, 45, 52, 79, 103, 114, 125, 190, 195, 203*
Hersh, Anita *50*
Hirst, Ric *41, 203*
Iris Color *74*
Livolsi, Michael *147*
Markis, Beverly *71*
McEnrue, Marianne *21*
Meyers, Herbert M. *14, 59, 160, 170, 236, 242*
Pennington, Alex *41*
Terrill, Paul *18*